METHODS IN MOLECULAR BIOLOGY™

Series Editor
John M. Walker
School of Life Sciences
University of Hertfordshire
Hatfield, Hertfordshire, AL10 9AB, UK

For further volumes:
http://www.springer.com/series/7651

Ancient DNA

Methods and Protocols

Edited by

Beth Shapiro

Department of Ecology and Evolutionary Biology, University of California Santa Cruz, A414 Earth & Marine Sciences, Santa Cruz, CA 95064, USA

Michael Hofreiter

Department of Biology, The University of York, Wentworth Way, Heslington, York YO10 5DD, UK

Humana Press

Editors
Beth Shapiro
Department of Ecology and Evolutionary Biology
University of California Santa Cruz
A414 Earth & Marine Sciences
Santa Cruz, CA 95064, USA
beth.shapiro@gmail.com

Michael Hofreiter
Department of Biology
The University of York
Wentworth Way, Heslington
York YO10 5DD, UK
michael.hofreiter@york.ac.uk

ISSN 1064-3745 e-ISSN 1940-6029
ISBN 978-1-4939-5874-0 ISBN 978-1-61779-516-9 (eBook)
DOI 10.1007/978-1-61779-516-9
Springer New York Dordrecht Heidelberg London

Softcover re-print of the Hardcover 1st edition 2012

Printed on acid-free paper

Humana Press is part of Springer Science+Business Media (www.springer.com)

Preface

Research in ancient DNA began more than 25 years ago with the publication of short mitochondrial DNA sequence fragments from the quagga, and extinct subspecies of the plains zebra. This publication was soon followed by a study reporting a 3.4 kilobase sequence of human nuclear DNA from an Egyptian mummy. Although today many researcher believe this later finding was the result of contamination with modern DNA, it nevertheless had substantial influence on the early phase of ancient DNA research. Despite the attention received by these early studies, research on ancient DNA only really gained momentum after the invention of the polymerase chain reaction, or PCR. This technology suddenly allowed millions of copies to be made of the few remaining ancient DNA molecules that in fortunate circumstances were preserved in fossils and museum specimens. In fact, without the invention of PCR, it is unlikely that ancient DNA research would ever have resulted in more than a few reports of short DNA fragments with little biological significance.

The use of PCR in ancient DNA research has been a double-edged sword. It has not only made possible many interesting studies, but has also facilitated the publication of some spectacularly wrong results. The best-known example of this is probably the publication of presumed dinosaur DNA sequences, which were later shown to be derived from modern human contamination. Presumed ancient DNA sequences were also reported from insects and plants embedded in pieces of amber and from water-logged plant fossils that were many millions of years old. Today, all of these are assumed to have been the result of contamination of samples, reagents, or experiments with modern DNA. These false positive results, which at the time were mostly published in high-profile journals, damaged the scientific reputation of the field, and it has taken many years to recover from this damage.

To some extent, these spectacular failures obscured the many sound, albeit less dazzling, studies that were published at the same time. The first Pleistocene-age DNA sequences from mammoth and cave bears were reported in 1994, and the first attempt to determine the phylogenetic position of the extinct moa within ratite birds was published in 1992. The potential of ancient DNA to investigate temporal changes in genetic diversity in populations was recognized even earlier: the first study, albeit only spanning a temporal period of approximately 70 years, was published in 1990. This was followed some years later by a study of European rabbits that extended the time frame for population genetics using ancient DNA to the Pleistocene/Holocene boundary, some 10,000 years ago.

For the next 10 years, the field of ancient DNA saw steady progress with regard to the age and type of samples used, the length of sequence analyzed, and the number of specimens included. In 2000, the first population study using Pleistocene-age DNA was published. This study, which focused on brown bears in Alaska, was important in that it showed that long-held beliefs regarding the evolution and establishment of modern phylogeographic patterns (the spatial structure of genetic diversity in a species) were incorrect. This work had a profound influence on the understanding of long-term population dynamics and dispersals during the Pleistocene and Holocene, and was followed by numerous studies showing that populations are far more dynamic units than previously assumed.

Only a year later, the first complete mitochondrial genomes of an extinct species were published independently by two research groups working on moa. These studies showed that despite the fragmented and damaged nature of ancient DNA molecules, it is possible to obtain longer DNA sequences from at least some ancient samples.

In parallel to the overall increase in length of the ancient DNA sequences obtained, the field also saw a significant increase in the age of the samples from which DNA sequences could be retrieved. Although, as noted above, all the extreme claims of millions of years old DNA were later shown to be false positives, the age of truly endogenous ancient DNA sequences was increasing considerably. The only authenticated ancient DNA sequences from the pre-PCR area, those of the quagga, were only 140 years old. Soon after PCR, maize sequences of about 1,000 years were reported in 1988, and by 1994, the oldest authentic DNA sequences dated to 40,000 years old. At the time of writing, the oldest published sequences come from a Greenland ice core and date to at least 500,000 years. Overall, over the lifetime of ancient DNA as a research field, the age of the investigated sequences has increased by more than four orders of magnitude.

Finally, the types of substrates used for ancient DNA extraction also have broadened tremendously. The first ancient DNA studies used soft tissue, building on the assumption that as these tissues, such as muscle, contain a lot of DNA in living organisms, they should also retain more DNA *postmortem* than other, less DNA-rich tissues. As for many assumptions made about ancient DNA, this proved to be false. The first ancient DNA sequences isolated from bone were reported in 1989, and, as it turned out, ancient bone contains on average much more DNA than ancient soft tissue, despite that in the living organism it contains much less DNA. Bone appears to preserve DNA much better than soft tissue, presumably because DNA adheres to the bone hydroxyl-apatite, and part of the DNA may even be preserved inside small hydroxyl-apatite crystals where it is protected from degradation. For almost 10 years, researchers concentrated mostly on bone as a source of ancient DNA, not only because it preserves DNA quite well, but also because it is rather abundant in the fossil record. In 1998, another, more unusual source of ancient DNA was opened up: coprolites, or subfossil faeces, which are found most often in cave sites in dry areas, especially in south-western North America. Since then, the variety of ancient DNA sources has increased steadily, with hair in 2001, packrat middens in 2002, sediment in 2003, feathers in 2009 and, most recently, eggshells in 2010. Thus, it is probably fair to say that most available substrates have by now been probed for ancient DNA and almost all yield DNA at least occasionally.

All the progress described above was mainly driven by the invention of and subsequent modifications to PCR. However, in 2005, a second revolution in ancient DNA research began with the introduction of the first of many so-called next-generation sequencing (NGS) technologies. The first generation of NGS machines resulted in an approximately 300-fold increase in DNA sequence throughput compared to traditional Sanger sequencing. Since then, DNA sequence throughput of NGS technologies has increased by another four orders of magnitude. Similar to PCR, these new technologies were rapidly adopted by the ancient DNA research community, and the first publication reporting ancient DNA sequences obtained by NGS was published only a few months after the technology itself had been published. Although this first publication was a mere proof-of-principle study, as it reported "only" 13 million base-pairs of mammoth nuclear DNA, it paved the way for more ambitious projects. Thus, in 2008, the first low-coverage (0.8-fold) draft genome of an extinct species, the mammoth, was published, and in 2010, the first high-coverage (20-fold) ancient human genome, obtained from the hair of a 4,000-year-old palaeo-eskimo was released. This was followed by 1.3- and 1.9-fold coverage genomes of Neanderthals and another, previously unrecognized hominid from Denisova Cave in Siberia.

NGS not only allows genomes to be sequenced from ancient remains. It has also resulted in the reconstruction of multiple, complete, ancient mitochondrial genomes, either via shotgun sequencing or in combination with multiplex PCR or hybridization capture approaches. Multiple (up to 30) complete or almost complete mitochondrial genomes have been obtained for cave bears, mammoths, and Neanderthals, and smaller numbers of mtDNA genomes have been obtained from ancient remains of other species including mastodon, short-faced bear, aurochs, Tasmanian tiger, and polar bear, and also from fossils of anatomically modern humans.

While the inventions of PCR and NGS clearly mark the two major revolutions in ancient DNA research thus far, progress has also been made in many smaller steps, including improved DNA extraction techniques, modifications to the PCR such as two-step multiplex PCR, and analytical approaches facilitating the analysis of time-structured data.

Progress in ancient DNA research has been inherently technology-driven. It may therefore come as a surprise that despite this importance of the appropriate methodological approaches in ancient research, no publication exists so far that summarizes current approaches toward the retrieval and analysis of ancient DNA sequences. This book attempts to close this gap. The chapters that follow describe a wide range of technologies, beginning with guidelines for the setup of an ancient DNA laboratory, describing extraction protocols for a wide range of different substrates and instructions for PCR and NGS library preparation, and finally suggesting appropriate analytical approaches in order to make sense of the sequences obtained. The chapters are written in a protocol-like style to make them accessible for every-day use in the lab. In addition, several chapters describe case studies linked to a protocol that illustrate what can actually be done using the described approaches. Due to these comprehensive but at the same time easily accessible protocols and illustrative case studies, we hope this book will be an interesting and useful source of information for the beginner and experienced researcher in ancient DNA alike.

We express our sincere thanks to all authors for their willingness to share their time and their trade secrets, and to Prof. John Walker at Humana Press for giving us the opportunity to assemble this collection of protocols.

Santa Cruz, CA, USA — ***Beth Shapiro***
York, UK — ***Michael Hofreiter***

Contents

Contributors

ADRIAN W. BRIGGS • *Department of Genetics, Harvard Medical School, 77 Avenue Louis Pasteur, Boston 02115, MA, USA*

ANDREW A. CLACK • *McMaster Ancient DNA Center, McMaster University, 1280 Main Street WestHamilton, ON, Canada L8S 4L9; Department of Biology, The Pennsylvania State University, 326 Mueller Laboratory,University Park, PA 16802, USA*

BETH SHAPIRO • *Department of Ecology and Evolutionary Biology, University of California Santa Cruz, A414 Earth & Marine Sciences, Santa Cruz, CA 95064, USA*

BRANDON LETTS • *Department of Biology, The Pennsylvania State University, 320 Mueller Laboratory,University Park, PA 16802, USA*

CHARLOTTE L. OSKAM • *Ancient DNA Laboratory, School of Biological Sciences and Biotechnology, Murdoch University, South Street, Perth 6150, WA, Australia*

ELMIRA MOHANDESAN • *Allan Wilson Centre for Molecular Ecology and Evolution, Institute of Natural Sciences, Massey University, Private Bag 102904 NSMC, Auckland, New Zealand*

GREGER LARSON • *Department of Archaeology, Durham University, South Road, Durham DH1 3LE, UK*

HENDRIK POINAR • *McMaster Ancient DNA Centre, McMaster University, Hamilton, ON, Canada*

JAMES HAILE • *Ancient DNA Laboratory, School of Biological Sciences and Biotechnology, Murdoch University, South Street, Perth 6150, WA, Australia*

LOGAN KISTLER • *Department of Anthropology, The Pennsylvania State University, 409 Carpenter Building, University Park, PA 16802, USA*

MARTIN KIRCHER • *Department of Evolutionary Genetics, Max Planck Institute for Evolutionary Anthropology, Deutscher Platz 6, D-04103 Leipzig, Germany*

MATTHIAS MEYER • *Max Planck Institute for Evolutionary Anthropology, Deutscher Platz 6, 04103 Leipzig, Germany*

MATHIAS STILLER • *Department of Biology, The Pennsylvania State University, 320 Mueller Laboratory, University Park, PA 16802, USA*

MELANIE KUCH • *McMaster Ancient DNA Centre, McMaster University, Hamilton, ON, Canada*

MICHAEL BUNCE • *Ancient DNA Laboratory, School of Biological Sciences and Biotechnology, Murdoch University, South Street, Perth 6150, WA, Australia*

MICHAEL HOFREITER • *Department of Biology, The University of York, Wentworth Way, Heslington, York YO10 5DD, UK*

MICHAEL KNAPP • *Allan Wilson Centre for Molecular Ecology and Evolution, Department of Anatomy and Structural Biology, University of Otago, Dunedin 9016, New Zealand*

MORTEN E. ALLENTOFT • *Ancient DNA Laboratory, School of Biological Sciences and Biotechnology, Murdoch University, South Street, Perth 6150, WA, Australia*
NADIN ROHLAND • *Department of Genetics, Harvard Medical School, 77 Avenue Louis Pasteur, Boston, MA 02115, USA*
PATRICIA HEYN • *Max Planck Institute of Molecular Cell Biology and Genetics, Pfotenhauerstrasse 108, 01307 Dresden, Germany*
PAULA F. CAMPOS • *Natural History Museum of Denmark, University of Copenhagen, Østervoldgade 5-71350, Copenhagen, Denmark*
ROSS BARNETT • *Department of Archaeology, Durham University, South Road, Durham DH1 3LE, UK*
ROSS D.E. MACPHEE • *American Museum of Natural History, New York, NY 10024, USA*
SIMON Y.W. HO • *School of Biological Sciences, University of Sydney, Sydney, 2006 NSW, Australia*
STEFAN PROST • *Allan Wilson Centre for Molecular Ecology and Evolution, Department of Anatomy and Structural Biology, University of Otago, Dunedin, New Zealand; Department of Integrative Biology, University of California, Berkeley, CA, USA*
SUSANNE HORN • *Max Planck Institute for Evolutionary Anthropology, Germany and German Cancer Research Center (DKFZ), Heidelberg, Germany*
STEPHEN M. WAGNER • *Department of Biology, The Pennsylvania State University, 320 Mueller Laboratory, University Park, PA 16802, USA*
TARA L. FULTON • *Department of Biology, The Pennsylvania State University, 320 Mueller Laboratory, University Park, PA 16802, USA*
THOMAS M.P. GILBERT • *Natural History Museum of Denmark, University of Copenhagen, Østervoldgade 5-7, DK 1350, Copenhagen, Denmark*

Chapter 1

Setting Up an Ancient DNA Laboratory

Tara L. Fulton

Abstract

Entering into the world of ancient DNA research is nontrivial. Because the DNA in most ancient specimens is degraded to some extent, the potential for contamination of ancient samples and DNA extracts with modern DNA is considerable. To minimize the risk associated with working with ancient DNA, experimental protocols specific to handling ancient specimens have been introduced. Here, I outline the challenges associated with working with ancient DNA and describe guidelines for setting up a new ancient DNA laboratory. I also discuss steps that can be taken at the sample collection and preparation stage to minimize the potential for contamination with exogenous sources of DNA.

Key words: Ancient DNA, aDNA, DNA damage, Laboratory setup, Contamination, Sub-sampling, Sample preparation, Guidelines

1. Introduction

The field of ancient DNA (aDNA) was born in 1984, when DNA sequences were successfully recovered from the extinct quagga, a relative of the zebra (1). With the advent of the polymerase chain reaction (PCR) (2), the field began to take shape (3) and has taken off during the last two decades. The power of aDNA is that it offers a window into the past that modern DNA or paleontological studies alone cannot provide. It has been widely adopted to address questions relating to, for example, the history and relationships of hominids (4), plant and animal domestication (5–8), population dynamics and diversity through time (9–13), and phylogenetics of extinct species (14–16). While aDNA can be a powerful tool, it is one that should be handled with caution.

Beth Shapiro and Michael Hofreiter (eds.), *Ancient DNA: Methods and Protocols*, Methods in Molecular Biology, vol. 840, DOI 10.1007/978-1-61779-516-9_1, © Springer Science+Business Media, LLC 2012

1.1. Difficulties of aDNA Work

1.1.1. Postmortem Degradation

DNA is frequently damaged while the organism is alive, but this damage is repaired via a suite of host repair mechanisms. DNA damage continues after death, but the repair pathways no longer function. As a result, few intact copies of aDNA tend to survive in old specimens, and those that remain are often highly fragmented and damaged (Table 1). Preservation in cold environments may slow or inhibit nuclease activity, reducing some of the damage that occurs immediately after death. However, environmental factors still work to cause DNA damage. Hydrolytic damage leads to single-strand breaks through direct cleavage or following depurination, fragmenting the DNA. Hydrolysis can also induce miscoding lesions, most commonly the deamination of cytosine to uracil, causing C–T transitions (17). Oxidation can induce lesions (17, 18) that block polymerases and either stop amplification or lead to "jumping PCR" (3) and the production of chimeric sequences.

Table 1
Types of ancient DNA damage

Damage type	Mechanisms	Effects	Solutions
Strand breaks	Nuclease activity Microorganism degradation Desiccation, heat, chemicals, etc. Direct cleavage (hydrolysis) Depurination causes a basic site (hydrolysis)	Low quantity of surviving DNA; short fragment length	Amplify short (<100–300 bp) overlapping fragments
Miscoding lesions via hydrolysis	*Deamination causes miscoding lesions* Adenine to hypoxanthine Cytosine to uracil[a] 5-methylcytosine to thymine[a] Guanine to xanthine	*Base misincorporations* A→G C→T C→T G→A	Multiple extractions and amplifications; cloning; UDG (uracil DNA glycosylase) to remove uracil
Blocking and Miscoding lesions via oxidation	*Base modifications* 5-OH-5-methylhydantoin (blocking) 5-OH-hydantoin (blocking) 8-oxoguanosine (miscoding G→T)	No amplification; jumping PCR Base misincorporation	Special polymerases; cloning; multiple amplifications
Crosslinks	DNA–DNA crosslinks via alkylation DNA–protein crosslinks (i.e., Maillard products)	No amplification	PTB (*N*-phenacylthiazolium bromide) to cleave crosslinks (but see (44))

[a]Generally, only C→T (or complementary strand G→A) transitions are observed (43)

Finally, crosslinks either within or between strands (18) will also block the polymerase.

Because of these types of damage, the majority of surviving DNA is generally short—less than 100 base pairs (bp) in length (19)—and contains damaged bases. The extent of this damage is highly sample-dependent and linked to preservation conditions (20). Cold, dry, temperature-stable environments such as permafrost regions and caves are among the best sources of well-preserved specimens and have permitted large-scale population studies (10, 11, 21). From these environments, reasonably well-preserved specimens with low levels of contamination have yielded exceptional amounts of data when next generation sequencing techniques are applied (22). As techniques improve, large-scale studies are progressively more attainable.

1.1.2. DNA Survival and Antediluvian DNA

Reports of antediluvian DNA (23), that is, sequences greater than one million years old, remain unsubstantiated and highly criticized. Particularly in the early days of aDNA research, such reports garnered much attention and publication in high-ranking journals. However, none of these claims have been independently substantiated through replication and many have been shown to be artifactual (for review (24–26)). The theoretical limit to DNA preservation remains between 100,000 and 1,000,000 years, although this varies considerably between preservation environments.

1.1.3. Contamination

The most serious complication of aDNA research stems from the small proportion of surviving copies of endogenous DNA in an extract, compared to the ubiquitous nature of environmental DNA. The high sensitivity of PCR allows amplification to proceed from only one or a few starting copies of the target sequence, but also often allows contaminating DNA to be amplified. Even if the level of contamination is extremely low, PCR will preferentially amplify modern DNA over damaged ancient molecules. For example, copies of the targeted fragment may contain blocking lesions or simply be in low abundance, so that it enters the exponential phase of the PCR many cycles after the reaction has begun. If only a few contaminant molecules are present and amplified during the initial cycles of the PCR, these will rapidly outnumber (and outcompete) any amplification of the aDNA.

Contamination may occur at many stages of processing an aDNA sample. The sample itself may be contaminated. For example, bones and teeth are porous, and contamination may occur via adherence or uptake of exogenous DNA, often from microorganisms residing in the depositional environment. Contamination may also occur during collection; this is a particular problem for human and microbial studies. Contamination may also be introduced during either the DNA extraction or amplification processes. Laboratory personnel may introduce their own DNA or any DNA carried into the lab, reagents may be contaminated with human or animal DNA (27), and airborne particulates may enter through the building air supply. Previously amplified DNA present in the laboratory environment is a particularly dangerous source of contaminating DNA. Even the tiny amount of DNA that is aerosolized when a tube is opened is likely to contain over a million copies of template in a volume as small as 0.005 μL. This is potentially thousands of times more copies than all the DNA that remains in an ancient sample (26). Therefore, it is crucial to maintain strict separation between the laboratory in which ancient samples are prepared and the post-PCR area.

Of course, not all samples share the same potential for contamination. Studies of ancient humans or microorganisms are at highest risk for contamination due to the pervasive nature of both potential contaminants. Ancient human sequences are not likely to differ substantially from modern humans, making identification of contaminants nearly impossible. Bacteria, and in particular environmental bacteria, are as yet so poorly characterized that any novel sequence isolated may simply represent an uncharacterized lineage. Thus, risk assessment at the outset of any aDNA project is critical, and study design must consider contamination potential as well as the information potential from the targeted data (28).

2. Guidelines for aDNA Research

Before discussing the guidelines that have been proposed for aDNA research, it is important to note that following these guidelines as a mere checklist will never guarantee that the sequences produced are authentic to the samples from which they were extracted. The burden is placed upon the researcher to critically analyze the project design, to assess which criteria are pertinent and, more importantly, to determine whether the results obtained from the experiment make sense, both in an evolutionary and experimental context (25, 28). Not every study must comply with all of the criteria presented below to be credible. For example, results from associated remains may not always be available or an assessment of biochemical preservation may not be necessary if the sample appears to be well preserved and the results are replicable and sensible. Above all, scientific judgment and rigor should prevail.

Building upon previous guidelines (3, 29), nine criteria for authenticity were set out roughly a decade ago by Cooper and Poinar (30):

1. *Physical isolation of the pre-PCR aDNA facility and strict maintenance of a "one-way" rule of movement up the concentration gradient*: All reagents and equipment must only move in the direction of the pre-PCR facility (aka the "clean" laboratory) to the post-PCR facility. In many labs, additional precautions are taken so that once laboratory personnel have entered any building in which PCR is performed, they can only reenter the pre-PCR facility after fully showering and changing clothes (see Note 1).
2. *Negative extraction and PCR controls*: Extraction and PCR controls containing no DNA (negative controls) must be carried out alongside the sample(s) at every step. Positive controls (those containing DNA that have been shown previously to be successful) should be avoided due to the risk of cross-contamination. If using a positive control cannot be avoided (e.g., if a problem with a component of the extraction or amplification is suspected), previously successful *ancient* specimens should be used in place of any modern specimens.
3. *Appropriate molecular behavior*: An inverse relationship should be observed between the length of the targeted PCR fragment and the strength of the amplification. If very long fragments are amplified as readily as are shorter fragments, it is likely that the product is a modern contaminant.
4. *Reproducibility*: Within-lab replication of PCR amplifications, overlapping PCR products, and amplifications from multiple DNA extractions must be consistent. If differences occur, more replicates should be performed.
5. *Cloning*: At minimum, a subset of PCR amplifications (i.e., 10%, including all unusual results) should generally be cloned to assess damage, detect nuclear mitochondrial insertions (often called numts), chimeric sequences from jumping PCR, and contaminants.
6. *Independent replication*: A second, independent laboratory should be able to replicate any results that are obtained. Ideally, the specimen would be divided upon collection and sent to two separate facilities to avoid transfer of one lab's potential contaminants to the other. However, if the original specimen is contaminated, this will be faithfully replicated in both facilities.
7. *Biochemical preservation*: An assessment of the likelihood of DNA preservation may be performed, for example a test of amino acid racemization ((31), but see (32)), mass spectrometry to determine the peptide to single amino acid ratio, bone histology, damage determination via gas chromatography or mass spectrometry, and bone porosity/density (25).

8. *Quantitation of starting material*: If very few surviving DNA molecules are present, stochasticity in the amount or type of damage present in the starting molecules of PCR, and PCR error in early cycles, may produce sequence errors that appear in the majority of clones (33). Thus, more than two amplifications and preferably multiple extractions should be performed.
9. *DNA from associated remains*: Particularly for high-risk studies such as those on humans, DNA preservation from animal or other remains associated with the find lends confidence that the site conditions are conducive to DNA preservation.

These criteria have been subsequently refined and expanded. Additional criteria include the following:

10. *Use of a "carrier DNA" negative*: Additional controls containing nonamplifiable "carrier DNA" should also be included. Sometimes contaminants may be in such low concentration that they bind to plasticware and do not amplify. However, the presence of any other DNA (such as the target DNA) may carry them through the reaction, resulting in amplification of the contaminant molecules in the PCR tubes containing the sample DNA but misleadingly clean negative controls (29). Including carrier DNA, such as nontarget DNA from a different source (see Note 2), with the sample may also help to allow amplification of very low copy target DNA.
11. *Time-dependent or preservation-dependent pattern of DNA damage and sequence diversity* (26, 34): Sequences isolated from badly preserved samples should be more damaged than better preserved samples (as assessed via cloning or high-throughput sequencing).
12. *Phylogenetic sense* (29) *or otherwise reasonable results* (28): Critical assessment of the sensibility of the results obtained from an aDNA experiment is an important aspect of aDNA research. Although the sequence may be expected to be different from any known sequence, the results should be reasonable based on the genomic regions targeted and the questions asked. If the sequence is highly divergent from any known sequence, that result should make sense given the taxon being studied and the data available for comparison. For example, BLAST searching should be used to ensure that the sequences are neither human nor environmental contaminants when another species is expected.

3. Setting Up an Ancient DNA Laboratory

3.1. Setting Up the Ancient DNA Workspace

Appropriate setup of the aDNA workspace is critical if contamination is to be avoided. The aDNA facility should be isolated from any location where PCR is routinely performed, preferably in a separate building that does not house any PCR labs. Ideally, the room will be positively pressurized, so that air does not flow in from the adjoining room/hallway when the door is opened. A laminar flow hood or glove box provides a clean space for PCR setup, even if work is performed in dead space (no air flow). Reagents and equipment should never be taken into the aDNA workspace from a post-PCR facility.

When planning the layout of an aDNA lab, it is important to consider what experimental protocols will be performed in that facility. This will help to identify the amount of space required and determine whether space needs to be allocated to large pieces of equipment, such as freezers or large centrifuges. How many people are anticipated to work in the facility at a time? Can the sample preparation and extraction area be physically separated from the PCR preparation area? Have future increases in storage needs for frozen, refrigerated, or room temperature items, including samples, been considered? As in any lab, a brightly lit, highly organized facility with very little clutter on the benches is more likely to create an atmosphere conducive to the careful, precise work that is required in aDNA research.

Because everything must be newly purchased for the aDNA workspace, it is often useful to envision a "walk-through" of the procedures that will be performed. Many items, such as paper, writing utensils, cleaning supplies, or glassware, are taken for granted in established labs. It is very inconvenient and sometimes quite difficult to temporarily delay a protocol, so preparation is key. Consider a solution that must be made. Is a pen and paper or calculator for recipe calculations available? Measuring devices (graduated cylinders, pipettes, etc.) of appropriate size for each ingredient, a scale that is sufficiently sensitive to measure the required dry reagents, weigh boats, and scoops, are only a few items that should be considered. How will the solution be mixed—is a stirrer or hot plate required? If so, are stir bars of an appropriate size available in the lab? How will the solution be pH'd, if necessary? Is an appropriate storage container for the final solution available? How will all of the materials be sterilized before and after use? The time it takes to ask these questions is more than made up for when a procedure can be performed smoothly without requiring a break for several days as a forgotten reagent or piece of equipment is shipped.

3.2. Reagents and Materials

One should assume that all reagents and tools are contaminated with human DNA, even those labeled as sterile. Although few products are certified to be DNA-free, many are certified as nuclease-free; using these is key so that what little aDNA remains in an extract is not degraded upon contact with nuclease-contaminated materials. For the greatest assurance that potential contaminants are removed, especially for human studies, all solutions should be filter sterilized.

All equipment must be decontaminated before use, for example by UV irradiation (45 W, 72 h), baking at 180°C for 12 h, acid-treatment with 2.5 M HCl for 48 h, or 3–5% bleach (up to 50% dilution of store bleach; sodium hypochlorite) for 48 h (26, 35). It has been recommended that autoclaving is not used for sterilization, as it breaks DNA into short (~100 bp) fragments and may increase the potential for bacterial contamination (35). However, many of the methods listed above may result in problems if they are not performed effectively. As all function to break down DNA, if the reaction is incomplete (i.e., not high enough chemical concentration; insufficient time of exposure to UV), the resulting fragmented, damaged, modern DNA appears and behaves as aDNA. Though widely used, UV destruction of DNA mimics that of natural environmental processes, causing photoproducts in DNA that lead primarily to C–T transitions (36). Bleach causes oxidative damage to DNA, producing chlorinated base products and cleaving DNA into progressively smaller fragments (37). If this process is incomplete, modern contaminant DNA will be fragmented and appear ancient. Thus, it is important to adhere fully to decontamination procedures, so as not to disguise modern contamination as the targeted aDNA.

Studies of domestic animals should also consider that PCR reagents are potentially contaminated not only by human DNA, but also by animal DNA (27). Thus, negative controls including carrier DNA are important in both human and animal studies, as is considering the source of the reagents. For example, if ancient cattle or bison DNA is the target, rabbit serum albumin (RSA) may be a better choice than bovine serum albumin (BSA) in PCR.

3.3. Maintaining a Sterile Space

To maintain a sterile working environment, all personnel should wear a full body suit including a hood, mask, shoe covers, and gloves at all times. Personnel should not enter the lab unless they have showered and changed into clothing that has not been in the PCR lab (see Note 1). Every surface should be cleaned before and after any work is performed; an additional thorough weekly cleaning is also important to maintain a sterile work environment. Surfaces can be cleaned in a number of ways to destroy any contaminant DNA present: 3–10% bleach followed by 70% ethanol (to clean away bleach and avoid corrosion), acid, and/or nightly UV irradiation. Keep in mind that the same problems of insufficient decontamination apply as discussed in the previous section.

3.4. Sample Preparation and Storage

The most effective way to avoid contamination of samples is to adhere to stringent aDNA sterility protocols from the moment that the sample is excavated onward. Contamination at the point of collection is a serious problem, particularly for bones and teeth (38, 39). However, it is not always possible to be certain of the collection and storage history of every specimen. Even those samples collected in the most sterile manner may still have surface contaminants from the depositional environment. Many protocols have been proposed that attempt to remove surface contaminants from ancient bones and teeth. These include the physical removal of the outer surface, washing or prolonged exposure of the outer surface to chemicals including water, EDTA, bleach, ethanol, acid, or hydrogen peroxide, UV irradiation of the sample, and/or extraction of the inner material (37). If the sample is well preserved, bleaching the bone powder can be effective to degrade contaminant DNA preferentially over endogenous DNA (40).

Ancient DNA has been obtained from many different kinds of substrates. Hair has recently shown excellent promise for the preservation of aDNA and exclusion of environmental contaminants (41), as was highlighted recently with the publication of the first ancient human genome sequenced from permafrost-preserved hair (42). However, because of their comparative abundance, bones and teeth remain the most commonly used substrates for aDNA work. When obtaining powder from bones or teeth for extraction (see Note 3), it is important to protect the area where PCR is set up, either by separating or enclosing the bone powdering area or by installing an air vent that will gently draw the bone powder out of the air and separately enclosing the PCR setup area in a flow hood or glove box.

The most appropriate protocol for long-term storage of ancient specimens varies depending on how the species were collected. If a sample was frozen upon collection, it is ideal to maintain that temperature. If a sample was collected at room temperature, it should be stored in a cool, dry environment but may not benefit from being frozen, in particular if several freeze/thaw cycles are anticipated. In general, simply avoiding environmental conditions that are known to promote DNA damage is key to sample preservation. A cool, dry, temperature-stable environment is ideal. Avoid heat, freeze/thaw cycles, and moisture. Although the depositional and preservation conditions are most important in the survival of DNA through time, poor treatment after sample collection can quickly degrade DNA that has persisted over thousands of years.

Work with aDNA is time-consuming and expensive. However, when care and appropriate precautions are taken from the outset, it can be a powerful tool for investigating evolutionary processes that cannot be addressed using modern data alone.

4. Notes

1. This also applies to notebooks, carrier bags, jackets, etc. It is practical to have a storage area outside the clean lab where personnel can leave their "PCR-contaminated" belongings.

2. Carrier DNA should be selected to be a type of DNA that will not be amplified in the PCR reaction. For example, lambda or vector DNA is easily accessible and makes good carrier DNA for mammalian studies.

3. A Dremel rotary tool with a cutting blade is useful for scraping off the surface of a bone, as well as for cutting out sections that can later be powdered using a bone mill. Other bits, such as a Dremel engraving cutter or a regular drill bit, are useful for hollowing out sections of bone underneath the surface, reducing the amount of damage that is visible as well as accessing the inner material that is less likely be contaminated or exposed to damaging agents.

References

1. Higuchi R, Bowman B, Freiberger M et al (1984) DNA-sequences from the Quagga, an extinct member of the horse family. Nature 312:282–284
2. Saiki RK, Scharf S, Faloona F et al (1985) Enzymatic amplification of beta-globin genomic sequences and restriction site analysis for diagnosis of sickle-cell anemia. Science 230:1350–1354
3. Pääbo S, Higuchi RG, Wilson AC (1989) Ancient DNA and the polymerase chain-reaction—the emerging field of molecular archaeology. J Biol Chem 264:9709–9712
4. Green RE, Krause J, Briggs AW et al (2010) A draft sequence of the Neandertal genome. Science 328:710–722
5. Edwards CJ, Bollongino R, Scheu A et al (2007) Mitochondrial DNA analysis shows a Near Eastern Neolithic origin for domestic cattle and no indication of domestication of European aurochs. Proc Biol Sci 274:1377–1385
6. Larson G, Liu RR, Zhao XB et al (2010) Patterns of East Asian pig domestication, migration, and turnover revealed by modern and ancient DNA. Proc Natl Acad Sci U S A 107:7686–7691
7. Leonard JA, Wayne RK, Wheeler J et al (2002) Ancient DNA evidence for Old World origin of New World dogs. Science 298:1613–1616
8. Goloubinoff P, Pääbo S, Wilson AC (1993) Evolution of Maize inferred from sequence diversity of an Adh2 gene segment from archaeological specimens. Proc Natl Acad Sci U S A 90:1997–2001
9. Stiller M, Baryshnikov G, Bocherens H et al (2010) Withering away-25,000 years of genetic decline preceded cave bear extinction. Mol Biol Evol 27:975–978
10. Shapiro B, Drummond AJ, Rambaut A et al (2004) Rise and fall of the Beringian steppe bison. Science 306:1561–1565
11. Campos PF, Willerslev E, Sher A et al (2010) Ancient DNA analyses exclude humans as the driving force behind late Pleistocene musk ox (*Ovibos moschatus*) population dynamics. Proc Natl Acad Sci U S A 107:5675–5680
12. Leonard JA, Wayne RK, Cooper A (2000) Population genetics of ice age brown bears. Proc Natl Acad Sci U S A 97:1651–1654
13. Pinsky ML, Newsome SD, Dickerson BR et al (2010) Dispersal provided resilience to range collapse in a marine mammal: insights from the past to inform conservation biology. Mol Ecol 19:2418–2429
14. Shapiro B, Sibthorpe D, Rambaut A et al (2002) Flight of the dodo. Science 295:1683
15. Orlando L, Metcalf JL, Alberdi MT et al (2009) Revising the recent evolutionary history of equids using ancient DNA. Proc Natl Acad Sci U S A 106:21754–21759
16. Krause J, Unger T, Nocon A et al (2008) Mitochondrial genomes reveal an explosive radiation of extinct and extant bears near the Miocene-Pliocene boundary. BMC Evol Biol 8:220
17. Lindahl T (1993) Instability and decay of the primary structure of DNA. Nature 362: 709–715
18. Pääbo S (1989) Ancient DNA—extraction, characterization, molecular-cloning, and enzymatic amplification. Proc Natl Acad Sci U S A 86:1939–1943
19. Poinar HN, Schwarz C, Qi J et al (2006) Metagenomics to paleogenomics: large-scale sequencing of mammoth DNA. Science 311: 392–394

20. Hoss M, Jaruga P, Zastawny TH et al (1996) DNA damage and DNA sequence retrieval from ancient tissues. Nucleic Acids Res 24:1304–1307
21. Rohland N, Pollack JL, Nagel D et al (2005) The population history of extant and extinct hyenas. Mol Biol Evol 22:2435–2443
22. Hofreiter M (2008) Long DNA sequences and large data sets: investigating the Quaternary via ancient DNA. Quat Sci Rev 27:2586–2592
23. Lindahl T (1993) Recovery of antediluvian DNA. Nature 365:700
24. Hofreiter M, Serre D, Poinar HN et al (2001) Ancient DNA. Nat Rev Genet 2:353–359
25. Pääbo S, Poinar H, Serre D et al (2004) Genetic analyses from ancient DNA. Annu Rev Genet 38:645–679
26. Willerslev E, Cooper A (2005) Ancient DNA. Proc Biol Sci 272:3–16
27. Leonard JA, Shanks O, Hofreiter M et al (2007) Animal DNA in PCR reagents plagues ancient DNA research. J Archaeol Sci 34:1361–1366
28. Gilbert MTP, Bandelt HJ, Hofreiter M et al (2005) Assessing ancient DNA studies. Trends Ecol Evol 20:541–544
29. Handt O, Hoss M, Krings M et al (1994) Ancient DNA—methodological challenges. Experientia 50:524–529
30. Cooper A, Poinar HN (2000) Ancient DNA: do it right or not at all. Science 289:1139
31. Poinar HN, Hoss M, Bada JL et al (1996) Amino acid racemization and the preservation of ancient DNA. Science 272:864–866
32. Collins MJ, Penkman KE, Rohland N et al (2009) Is amino acid racemization a useful tool for screening for ancient DNA in bone? Proc Biol Sci 276:2971–2977
33. Handt O, Krings M, Ward RH et al (1996) The retrieval of ancient human DNA sequences. Am J Hum Genet 59:368–376
34. Hebsgaard MB, Phillips MJ, Willerslev E (2005) Geologically ancient DNA: fact or artefact? Trends Microbiol 13:212–220
35. Willerslev E, Hansen AJ, Poinar HN (2004) Isolation of nucleic acids and cultures from fossil ice and permafrost. Trends Ecol Evol 19:141–147
36. Griffiths AJF (2005) Introduction to genetic analysis, 8th edn. W.H. Freeman, New York
37. Kemp BM, Smith DG (2005) Use of bleach to eliminate contaminating DNA from the surface of bones and teeth. Forensic Sci Int 154: 53–61
38. Gilbert MTP, Hansen AJ, Willerslev E et al (2006) Insights into the processes behind the contamination of degraded human teeth and bone samples with exogenous sources of DNA. Int J Osteoarchaeol 16:156–164
39. Sampietro ML, Gilbert MTP, Lao O et al (2006) Tracking down human contamination in ancient human teeth. Mol Biol Evol 23: 1801–1807
40. Salamon M, Tuross N, Arensburg B et al (2005) Relatively well preserved DNA is present in the crystal aggregates of fossil bones. Proc Natl Acad Sci U S A 102:13783–13788
41. Gilbert MTP, Menez L, Janaway RC et al (2006) Resistance of degraded hair shafts to contaminant DNA. Forensic Sci Int 156: 208–212
42. Rasmussen M, Li YR, Lindgreen S et al (2010) Ancient human genome sequence of an extinct Palaeo-Eskimo. Nature 463:757–762
43. Hofreiter M, Jaenicke V, Serre D et al (2001) DNA sequences from multiple amplifications reveal artifacts induced by cytosine deamination in ancient DNA. Nucleic Acids Res 29:4793–4799
44. Rohland N, Hofreiter M (2007) Comparison and optimization of ancient DNA extraction. Biotechniques 42:343–352

Chapter 2

A Phenol–Chloroform Protocol for Extracting DNA from Ancient Samples

Ross Barnett and Greger Larson

Abstract

The myriad downstream applications of ancient DNA (aDNA) analysis all ultimately require that sequence data are generated from extracts of ancient material. DNA extraction from tissues known to contain preserved biomolecules (e.g. teeth, hair, tissue, bone) relies on subtle modifications of a basic technique that has been in use for nearly two decades. Multiple DNA extraction protocols have been introduced, with varying levels of success depending on tissue type and the long-term preservation environment to which the ancient tissue was exposed. Here, we describe the phenol–chloroform method for extracting aDNA from any tissue type. This commonly employed method allows for the recovery of total nucleic acid content with minimal loss of low molecular weight double-stranded DNA.

Key words: Ancient DNA, Extraction, Bone, Teeth, Hair, Tissue, Phenol, Chloroform

1. Introduction

Over the past quarter century, there has been an enormous increase in the use of genetic data extracted from the preserved remains of plants and animals in questions of evolution, taxonomy, and population genetics. To extract DNA from ancient tissues, most of these studies have relied on simple modifications of a DNA extraction technique that was used in the late 1980s to generate the first ancient genetic data (1, 2). This technique, which has been known as the phenol–chloroform technique, has stayed in favour due to its ease of application and ability to harvest complete nucleic acid fractions. Like all DNA extraction techniques using aDNA, working in a sterile environment and taking careful measures to prevent contamination of the ancient samples and extracts throughout the protocol are crucial in the successful extraction of endogenous genetic material (3, 4).

Beth Shapiro and Michael Hofreiter (eds.), *Ancient DNA: Methods and Protocols*, Methods in Molecular Biology, vol. 840,
DOI 10.1007/978-1-61779-516-9_2,

The phenol–chloroform method relies on an initial digestion step to break down crystalline minerals, proteins, and complex lipids present in the sample. Further processing sequesters the nucleic acid fraction from other components by repeated separation into hydrophobic and aqueous phases. Final passing through a membrane filter concentrates the DNA in the sample. The final solution of total DNA can be used for PCR, next-generation sequencing, and other applications.

2. Materials

All reagents, consumables, and equipment used throughout this protocol should be sterilised, either through purchase or through processing with UV irradiation or cleaning with bleach solution and ethanol (5–7). Plasticware should be sterile, single-use, and preferably designed to prevent cross-contamination (e.g. pipette tips with aerosol barrier). All solutions should be prepared using ultrapure water (18.2 MΩ at 25°C).

2.1. Sample Preparation

1. Hand-held drill with disposable abrasive discs.
2. Sodium hypochlorite (bleach, 10–20% solution).
3. Ethanol (95–100%).
4. Freezer mill, shaker mill, or other device for grinding samples into powder (see Note 1).
5. Aluminium foil.
6. *N* prepared sterile plastic tubes (15 mL) (see Note 2).

2.2. Chelation

1. EDTA chelation buffer: Ethylenediaminetetraacetic acid (EDTA) 1 M solution pH 8.0 (see Note 3).
2. Rotary mixer, wheel or similar device to keep samples constantly in motion during incubation steps, suitable for use with 15-mL tubes.
3. Centrifuge suitable for use with 15-mL tubes.

2.3. Digestion

1. 1× Buffer:15 mM Tris–HCl (pH 8.0), 2.5 mM *N*-phenacylthiazone bromide (see Note 4).
2. 10×: Sodium dodecyl sulphate (Fisher) 10%w/v (see Note 5).
3. 10×: 25 mg mL^{-1} Proteinase K (see Note 6).
4. 10×: 500 mM Dithiothreitol (DTT) (see Note 7).
5. Laboratory incubator large enough to accommodate rotator.
6. Rotary mixer, wheel or similar device to keep samples constantly in motion during incubation steps, suitable for use with 15-mL tubes.

2.4. Phase Separation

1. 2*N* pre-prepared 15-mL phase separating tubes (light gel see Note 8) containing 6 mL of saturated pH 6.6 Phenol (see Note 9).
2. *N* pre-prepared 15-mL sterile plastic tubes containing 6 mL of chloroform (see Note 10).
3. Rotary mixer, wheel or similar device to keep samples constantly in motion during incubation steps, suitable for use with 15-mL tubes.
4. Centrifuge suitable for use with 15-mL tubes.

2.5. Concentration

1. *N* labelled micro-concentrators with a nominal molecular weight limit of 30 kDa and able to process 6 mL of solution (see Note 11).
2. Centrifuge suitable for use with 15-mL tubes.
3. *N* labelled sterile plastic tubes (1.5 mL).

3. Methods

As the method described below attempts to extract degraded, damaged DNA from samples that may be anywhere up to several hundred thousands of years old, it is necessary to work in an isolated aDNA workspace, ideally one that is isolated from normal molecular biology (especially PCR) work (4).

3.1. Sample Preparation

1. Prepare the work area by sterilising surfaces. First wash all surfaces with bleach and then rinse with ethanol. Wait until the surfaces dry completely before proceeding. A fume hood with integrated extraction fans is an ideal location for this protocol to be performed. Prepare several layers of aluminium foil to collect powdered sample as it is produced.
2. Thoroughly abrade the external surface of the bone/tooth sample using a hand-held drill with disposable cutting discs or equivalent (see Note 12). Discard the resulting powder, for example by collecting it in the upper layer of aluminium foil and discarding the foil and powder.
3. Reduce the bone/tooth section to powder using a shaker mill, freezer mill or similar device (see Note 13). The speed and other conditions of the powdering device should be adjusted to suit the mineralisation state of the sample. Generate as fine a powder as possible to maximise the surface area of the sample that will eventually contact the chelation solution (6, 8).
4. Collect the powder and transfer it to a labelled, sterile tube (15 mL).

3.2. Chelation

1. Add 15 mL of 1× chelation buffer to the powdered sample (see Note 14).
2. Add 15 mL of 1× chelation buffer to a labelled tube that does not contain any powder. This will be the negative extraction control.
3. Be sure that the powder and chelation buffer are well mixed. Place all 15-mL tubes on the rotary mixer and rotate overnight at room temperature.
4. Concentrate the samples and negative control by centrifugation at 4,000×g for 10 min or until all organic content has pelleted at the bottom of the tube.
5. Remove eluate and retain pellet (see Note 15).

3.3. Digestion

1. Prepare 4.2 mL of the 1× digestion buffer for each sample and add to this each tube containing an organic pellet using.
2. Add 0.6 mL of the 10× solution of SDS to each sample.
3. Add 0.6 mL of the 10× solution of proteinase K to each sample.
4. Add 0.6 mL of the 10× solution of DTT to each sample.
5. Place all 15-mL tubes on the rotary mixer. Place the rotator in the oven and rotate overnight at 55°C (see Note 16).

3.4. Phase Separation

1. Decant each digested sample into a corresponding pre-prepared tube containing phenol.
2. Place all 15-mL tubes on the rotary mixer and rotate at room temperature for 10 min
3. Centrifuge at 8,000×g for 10 min. The two phases will separate. If phase-lock or phase-divider tubes are used, the gel should have formed a barrier between the aqueous and hydrophobic layers (see Note 17). Decant the aqueous layer into the second, pre-prepared tube containing phenol.
4. Place all 15-mL tubes on the rotary mixer and rotate at room temperature for 10 min.
5. Centrifuge at 8,000×g for 10 min. As before, the two phases will separate. Decant the aqueous layer into a pre-prepared chloroform tube.
6. Place all 15-mL tubes on the rotary mixer and rotate at room temperature for 5 min.
7. Centrifuge at 8,000×g for 5 min. The two phases will separate.

3.5. Concentration

1. Carefully transfer the aqueous layer by pipette to a micro-concentrator (30 kDa membrane) (see Note 18).
2. Centrifuge at 8,000×g until the sample has completely passed through the membrane. Discard the filtrate.

3. Add 5 mL of ultrapure water to each sample and centrifuge again at 8,000×*g* until resolution into a final solution of 100–200 μL (see Note 19). The eluate will now contain your DNA.
4. Transfer the remaining solution by pipette to a sterile storage tube (e.g. 1.5 mL, see Note 20).
5. Store the DNA extract at −20°C (see Note 21).

4. Notes

1. Shaker mills, freezer mills, and other similar devices use friction to reduce samples to powder. The increase in surface area that results from powdering the sample allows for more efficient digestion.
2. Prior to beginning the protocol, prepare *N* sufficiently labelled tubes, where *N* is the number of samples plus negative controls (i.e. if 7 samples and one negative control are to be extracted, prepare $N=8$ labelled tubes, $2N=16$ phenol tubes, etc.).
3. EDTA is a strong chelator that is able to bind metallic ions such as Ca^{2+} and Mg^{2+} that are released during digestion.
4. Some experiments have shown that *N*-phenacylthiazone bromide may be useful in freeing DNA that has been chemically cross-linked to other biomolecules through diagenetic processes (9, 10).
5. SDS is a detergent that allows the solubilisation of lipids present in biological samples and denatures proteins. 10× SDS solution should be stored in the refrigerator. At low temperature, SDS can precipitate out of solution. If this occurs, place it in a warm oven for 5 min until the detergent has resolubilised.
6. Proteinase K is a protease that cleaves proteins, reducing them to their constituent amino acids. Proteinase K should be stored in the freezer where it will remain stable for several months.
7. DTT is a reducing agent that can cleave cystine–cystine bridges and disrupt the tertiary structure of some proteins, prior to digestion. DTT should be stored in the freezer where it will remain stable for several months.
8. Phase-lock or phase-dividing gels are useful in this capacity as they allow easy decanting of the aqueous phase. The inert gel can be added to tubes, and upon centrifugation, forms a barrier between the two phases. In the absence of gel, the aqueous phase can be transferred by careful manual pipetting.
9. Phenol is dangerous. Extreme caution must be exercised when aliquotting and transferring phenolic samples. Familiarise yourself with the appropriate safety information and correct disposal

methods before use. Polyethylene glycol should be kept to hand wherever phenol is used and those performing the technique should acquaint themselves with emergency procedures in case of phenol spills.

10. Chloroform is hazardous and should be handled with caution.
11. Concentrators use a membrane barrier of pre-determined pore size to prevent the passage of molecules larger than a certain molecular weight. A pore size with a molecular weight cut-off of 30,000 Da should prevent the loss of any nucleic acid larger than about 50 bp in length, but will allow removal of almost all other digested biomolecules.
12. This step removes any preservative coatings and potentially adsorbed environmental contaminants. Recent work suggests that the speed setting of the drill and the amount of friction produced can have a negative effect on DNA recovery (11). Generally, use the lowest speed setting possible for abrasion and cutting of bone and tooth.
13. Some of the sample should be retained for further extractions. Replication, either internal or external, may be necessary for ancient samples.
14. To adapt the protocol for tissue/hair/nail, the overnight chelation stage can be omitted. Begin at the digestion stage and add 0.5 mL of 1× Chelation buffer to the digestion buffer, SDS, proteinase K, and DTT (12). Follow the remainder of the protocol as described.
15. The eluate is also likely to contain DNA and can be retained for processing, either in parallel or at a later date (8).
16. The temperature setting for digestion can be modified depending on the sample. Recent work suggests that lower temperatures may have a beneficial effect on DNA recovery. If the temperature is lowered, increase the length of time the samples are left to rotate until complete digestion is achieved (8).
17. Very occasionally, the phase-lock tubes may not separate properly between aqueous and hydrophobic phases. In this case, care should be taken when manually removing the aqueous phase by pipette.
18. Adding a small volume of ultrapure water to the filters prior to adding the extract may aid in absorption of DNA to the membrane. Depending on the volume of extract to be processed, this step may have to be repeated multiple times until the entire sample has passed through the membrane.
19. Flushing water through the membrane after the entire sample has been passed through may help to further remove any potential inhibitors from the final extract.
20. After extraction, DNA can be roughly quantified by measurement on a spectrophotometric platform. Note that this does

not give an indication of how much of the DNA in the extract is derived from the sample vs. from co-extracted environmental contaminants.

21. It may be useful to subdivide the final extracts into aliquots of 20–50 μL and to use these as necessary. DNA is susceptible to damage from repeat freeze–thaw cycles (13) and should be defrosted as infrequently as possible.

Acknowledgments

Thanks to Beth Shapiro for the opportunity to contribute this chapter.

References

1. Hagelberg E, Clegg JB (1991) Isolation and characterization of DNA from archaeological bone. Proc Biol Sci 244:45–50
2. Hagelberg E, Sykes B, Hedges R (1989) Ancient bone DNA amplified. Nature 342:485
3. Hofreiter M, Serre D, Poinar HN, Kuch M, Pääbo S (2001) Ancient DNA. Nat Rev Genet 2:353–359
4. Cooper A, Poinar HN (2000) Ancient DNA: do it right or not at all. Science 289:1139
5. Leonard JA, Shanks O, Hofreiter M, Kreuz E, Hodges L, Ream W, Wayne RK, Fleischer RC (2007) Animal DNA in PCR reagents plagues ancient DNA research. J Archaeol Sci 34:1361–1366
6. Rohland N, Hofreiter M (2007) Ancient DNA extraction from bones and teeth. Nat Protoc 2:1756–1762
7. Handt O, Höss M, Krings M, Pääbo S (1994) Ancient DNA: methodological challenges. Experientia 50:524–527
8. Rohland N, Hofreiter M (2007) Comparison and optimization of ancient DNA extraction. Biotechniques 42:343–352
9. Poinar HN (2002) The genetic secrets some fossils hold. Acc Chem Res 35:676–684
10. Vasan S, Zhang X, Kapurnitou A, Bernhagen J, Teichberg S, Basgen J, Wagle D, Shih D, Terlecky I, Bucala R, Cerami A, Egan J, Uhlrich P (1996) An agent cleaving glucose-derived protein crosslinks in vitro and in vivo. Nature 382:275–278
11. Adler CJ, Haak W, Donlon D, Cooper A (2010) Survival and recovery of DNA from ancient teeth and bones. J Archaeol Sci 38(5):956–964
12. Gilbert MTP, Wilson AS, Bunce M, Hansen AJ, Willerslev E, Shapiro B, Higham TFG, Richards MP, O'Connell TC, Tobin DJ, Janaway RC, Cooper A (2004) Ancient mitochondrial DNA from hair. Curr Biol 14:R463–R464
13. Lindahl T (1993) Instability and decay of the primary structure of DNA. Nature 362: 709–715

Chapter 3

DNA Extraction of Ancient Animal Hard Tissue Samples via Adsorption to Silica Particles

Nadin Rohland

Abstract

A large number of subfossil and more recent skeletal remains, many of which are stored in museums and private collections, are potentially accessible for DNA sequence analysis. In order to extract the small amount of DNA preserved in these specimens, an efficient DNA release and purification method is required. In this chapter, I describe an efficient and straightforward purification and concentration method that uses DNA adsorption to a solid surface of silica particles. Comparative analysis of extraction methods has shown that this method works reliably for ancient as well as younger, museum-preserved specimens.

Key words: Ancient DNA, DNA extraction, Bones, Teeth, Museum-specimen, Silica, Column

1. Introduction

The most abundant faunal remains are partial skeletons. Bones and teeth are the hardest tissues of vertebrates and can persist for hundreds of thousands of years without fossilization if sediments or permafrost shield them from unstable environmental conditions. When environmental conditions are unfavorable for microbial life that would otherwise metabolize the hard tissue, this can lead to the preservation of DNA molecules within these ancient skeletons. Such conditions are common to permafrost regions, where large numbers of preserved faunal remains have been found. In more moderate climatic ecosystems, well-preserved skeletal remains can be found within sediment deposits in natural shelters such as caves.

Three major obstacles impede DNA analyses of ancient skeletal remains. First, the total amount of DNA preserved in very old bones and teeth is likely to be very small, and often the DNA fragments that do remain are highly damaged (1). The same may be

Beth Shapiro and Michael Hofreiter (eds.), *Ancient DNA: Methods and Protocols*, Methods in Molecular Biology, vol. 840,
DOI 10.1007/978-1-61779-516-9_3,

true for modern specimens that have been treated with chemical preservatives to prepare them for long-term storage in museums. Second, if DNA is preserved in ancient bones or teeth, it is often contaminated with DNA from bacteria, fungi, or other microbial organisms (2). Third, regardless of the environmental condition from which the sample is excavated, contaminating organic and inorganic compounds, such as humic acid and salts leaking from the surrounding soil, can accumulate in the cavities of these samples over the years. These are often coextracted along with the endogenous DNA of the sample. Therefore, ancient DNA extraction methods need not only to recover DNA molecules preserved in the samples efficiently, but also to remove contaminating compounds that may inhibit subsequent enzymatic reactions.

The solid matrix of bones and teeth promotes their physical preservation and the preservation of biomolecules within them. However, this matrix needs to be disrupted during the extraction process in order to release the DNA molecules into an aqueous solution so that it can be purified. Several DNA purification and concentration methods are used for ancient animal hard tissue samples. The purification method described here is a two-part process, where DNA is first adsorbed to the surface of silica particles and then salts and other contaminating chemicals are removed. The method is identical in concept and very similar in approach to methods employed in various commercially available kits.

In previous comparative analyses, we found DNA purification by adsorption to silica particles in suspension to perform best with respect to amplifiable DNA recovery from ancient bone and tooth samples when guanidinium isothiocyanate (GuSCN) was used as a chaotropic salt to drive the adsorption of DNA (3). GuSCN seems to prevent silica particles from adsorbing potentially inhibiting coextracts that may have accumulated in the samples. One advantage to using a solid phase to pull down the DNA from an aqueous solution is that the particles can be immobilized in an appropriate device. If these device(s) allow parallel processing, salt and other chemicals can easily be washed away and DNA eluted from many samples in parallel. Single column devices are commercially available, and using a vacuum device or a microcentrifuge to remove the buffers in between the steps allows for a moderate throughput for DNA extraction of ancient and historical museum samples (4).

The following protocol is presented using column devices and a vacuum manifold. If no vacuum manifold is accessible, the extraction can be performed using the columns and a microcentrifuge. It is also possible to perform the extraction without the columns by using regular 1.5- or 2.0-mL tubes and resuspension of the silica particles followed by centrifugation, rather than the simpler method (vacuum-mediated washing by flow-through) described below. However, it should be noted that some DNA may be lost as it adheres to the inside surface of the pipette tips during repeated resuspension steps.

The presence of intact cellular structures depends on the degradation state of the sample. A detergent and a reducing agent are recommended for more recent samples (4) and for well-preserved ancient samples such as those from permafrost environments. However, it is usually not necessary to use these when working with ancient specimens (5). Nevertheless, no negative effect has been observed when the detergents and reducing agents used below are included in the extraction buffer, even for very old, nonpermafrost specimens (3). These are therefore included in the extraction buffer described below.

2. Materials

HPLC grade water or water with a similar purity grade is recommended to prepare all solutions and suspensions.

2.1. DNA Release from Bony Specimen

1. Extraction buffer: 0.45 M EDTA (pH 8.0), 1% Triton-X 100, 50 mM DL-Dithiothreitol, 0.25 mg/mL proteinase K (see Note 1).
2. Cutting or drilling tool with exchangeable disposable bits or discs.
3. Mortar and pestle or freezer mill (e.g., SPEX SamplePrep 6750 Freezer/Mill; liquid nitrogen is needed) for grinding sample pieces into fine powder.
4. 15-mL tubes.
5. Rotary mixer, wheel, or similar device to keep samples constantly in motion during incubation steps.

2.2. DNA Purification and Concentration

1. Silicon dioxide (see Note 2).
2. 30% HCl.
3. Binding buffer: 5 M Guanidinium thiocyanate, 0.3 M sodium acetate (pH 5.2). Store in the dark (see Note 3).
4. Washing buffer: 50% Ethanol, 125 mM NaCl, 10 mM Tris–HCl, 1 mM EDTA (pH 8.0) (see Note 4)
5. Elution buffer: 10 mM Tris–HCl, 1 mM EDTA (pH 8.0).
6. 50-mL tubes.
7. 50-mL disposable serological pipettes.
8. Centrifuge capable of holding 15-mL tubes and reaching centrifugal force of 5,000 × *g*.
9. Columns (e.g. MobiCol "Classic," MobiTec, catalog number: M1003).
10. Filter (Filter (large) 10 μm pore size, MobiTec, catalog number: M2210).

11. Filter with 1 μm pore size (e.g., glass microfiber binder free Grade GF/B: 1 μm, Whatman, catalog number: 1821–070).
12. Hole punch with 7 mm diameter.
13. Forceps.
14. Centrifuge capable of holding 2.0-mL tubes and reaching centrifugal force of 16,000 × *g*.
15. Vacuum manifold and vacuum pump.
16. Collection tubes without lids.
17. Disposable VacConnectors (Qiagen, catalog number: 19407).
18. 1.5-mL tubes (see Note 5).

3. Methods

All steps are to be carried out at room temperature.

3.1. Preparing the Silica Suspension

1. Weigh 4.8 g of silicon dioxide into a 50-mL tube, add water to bring the mixture to 40 mL, and vortex extensively.
2. Let large particles settle down for 1 h.
3. Transfer 39 mL from the top of the solution into fresh 50-mL tube and let the solution settle for an additional 4 h.
4. Discard 35 mL from the top of the solution and add 48 μL of 30% HCl to the 4 mL pellet that remains.
5. Vortex, aliquot, and store the silica suspension at room temperature in the dark (see Note 6).

3.2. Preparing the Columns

1. Use forceps to place a large filter with 10 μm pore size in the column. Move the filter to the bottom of the column using the filter insertion tool provided with the filter.
2. Use a hole punch to make a smaller "fine filter" from the filter paper with 1 μm pore size.
3. Using the forceps, place the "fine filter" in the columns, and move it on top of the larger filter using the insertion tool.

3.3. Sample Preparation and DNA Release

1. After removing the surface of the sample with a fresh drilling bit at slow speed, drill into the densest part of the bone or the tooth root. Collect the powder. If a cutting tool is used instead of a drill, remove a compact part of the bone or the tooth root (again after removing the sample surface with a single-use cutting disc or blade). Grind the pieces of sample to as a fine powder as possible using mortar and pestle or a freezer mill. Collect approximately 250 mg of powder per sample into separate 15-mL tubes (see Note 7).

2. Add 5 mL of extraction buffer to each sample. Seal the tubes and incubate them for 16–24 h under constant agitation in the dark (see Note 8).

3.4. DNA Adsorption to Silica, Washing Steps, and Elution

1. Centrifuge the samples for 2 min at 5,000 × *g* and transfer as much of the liquid as possible into new 15-mL tubes.
2. Add 2.5 mL of binding buffer and 100 μL of well-mixed silica suspension to the extraction buffer in each tube. Incubate for 3 h in the dark under constant agitation (see Notes 9–11).
3. Place a disposable VacConnector onto the luer adapter of the vacuum manifold, then place the assembled column onto the VacConnector (depending on the manifold used, up to 24 columns can be handled in parallel).
4. Centrifuge the sample for 2 min at 5,000 × *g*, discard the supernatant, and resuspend the silica pellet in 400 μL of binding buffer. Transfer the suspension to the column and apply the vacuum (see Notes 12–14).
5. Place the column in a collection tube and centrifuge for 30 s at 16,000 × *g* (see Note 15).
6. Place the column back onto the VacConnector of the vacuum manifold. Add 450 μL of washing buffer to the column and apply the vacuum (see Note 16).
7. Repeat the washing step at least once while the column remains on the vacuum manifold (see Note 17).
8. Insert the column into a collection tube and centrifuge for 30 s at 16,000 × *g* (see Note 18).
9. Insert the column into a new, labeled 1.5-mL tube and allow the silica to air-dry by incubating the columns with open lids for about 3 min (see Notes 5 and 18).
10. Add 50 μL of elution buffer onto the center of the silica pellet and incubate the columns for 10 min with closed lids (see Notes 19 and 20).
11. Centrifuge for 1 min at 16,000 × *g* (see Note 21).

4. Notes

1. Always prepare the extraction buffer immediately before beginning the extraction, as proteinase K loses activity rapidly.
2. Recommended silicon dioxide: Sigma-Aldrich, catalog number: S5631.
3. The binding buffer is stable for at least 1 month. This buffer should be stored in the dark.

4. The washing buffer is stable for several months.
5. Low retention or siliconized tubes are recommended, which reduce DNA loss due to tube wall effects.
6. The silica suspension is stable for at least 1 month.
7. Do not exceed 250 mg/5 mL extraction buffer. It is possible to proportionally scale the extraction up or down when more or less sample material is used; use 1 mL/50 mg. It is crucial to adjust the binding buffer volume accordingly (see Notes 10 and 11).
8. Incubation can also be performed at 37°C, where proteinase K is more active than it is at room temperature. This may increase the DNA quantity, especially for younger samples with intact cell structures. Although increasing incubation time was not seen to have an influence in our test series of ancient samples (3), incubation time can be extended in order to completely digest the material; this may also increase the quantity of extracted DNA (6).
9. Silica must be extensively vortexed before adding to the extraction and binding buffer as particles quickly settle down.
10. If more or less extraction buffer was used, adjust the volume of binding buffer accordingly so that the ratio of extraction to binding buffer is 2:1.
11. The volume of silica suspension should also be adjusted proportionally when different extraction buffer volumes are used. The volume of silica suspension should be at least 50 μL, as too few silica particles may result in a loss of DNA molecules. If a very large volume of extraction buffer is used, do not exceed 200 μL of silica suspension per extraction/column, as adding more per column may result in incomplete washing and elution performance; instead, concentrate the extraction buffer prior to the adsorption step using appropriate filter systems (e.g., (2)), or distribute the silica over several columns when more than 200 μL of silica is used. However, the latter will result in higher elution volumes of less concentrated extract.
12. It is recommended that you keep the supernatant until the positive control gives satisfying results. If the extract of the positive control does not contain any DNA, you may repeat the adsorption and purification steps by adding freshly made silica suspension and proceed from the 3-h incubation step onwards.
13. If no columns are used, transfer the silica suspension into a 1.5- or 2.0-mL tube and perform the washing steps by resuspending the silica with washing buffer by pipetting. Then centrifuge for 30 s at 16,000 × *g* to pelletize the silica, and discard washing buffer by pipetting it off. Dry the silica for at least 10 min and resuspend the silica in 50 μL elution buffer by pipetting.

After final incubation for 10 min, centrifuge for 1 min at 16,000 × *g* and pipette off the extract into a fresh-labeled tube.

14. If you are not using a vacuum manifold, this step and all following washing steps can be performed using a microcentrifuge and collection tubes. For an even distribution of the silica particles over the filter and subsequently efficient washing performance, short, slow-speed centrifugation is recommended, followed by a 180° rotation of the column and another short, slow-speed centrifugation step after the silica is applied to the columns.
15. This is a crucial step to remove remaining salts and other chemicals, as remaining GuSCN can lead to incomplete elution of the DNA from the silica and/or inhibit subsequent enzymatic reactions.
16. Fresh VacConnectors are recommended.
17. If the silica is still deeply colored after two washing steps, it is possible to wash the silica with 450 μL binding buffer, followed by centrifugation and at least two washing steps with washing buffer. Washing with binding buffer seems to reduce the amount of colored and potentially inhibiting coextracted contaminants.
18. This is a crucial step to remove any salt and ethanol remains, which may lead to incomplete elution and/or inhibition of enzymatic reactions that follow.
19. If the silica particles are not evenly distributed on the filters, add the elution buffer on top of the thickest part of silica particles.
20. If more than 100 μL silica was used for adsorption, proportional increase of the elution buffer volume is recommended.
21. The elution step may be repeated. However, this increases the volume of extract, but also reduces the concentration of DNA in the extract.

Acknowledgments

I would like to thank Michael Hofreiter and Elizabeth Fels for linguistic improvements of the manuscript.

References

1. Pääbo S, Poinar H, Serre D, Jaenicke-Despres V, Hebler J, Rohland N, Kuch M, Krause J, Vigilant L, Hofreiter M (2004) Genetic analyses from ancient DNA. Annu Rev Genet 38:645–679
2. Noonan JP, Hofreiter M, Smith D, Priest JR, Rohland N, Rabeder G, Krause J, Detter JC, Pääbo S, Rubin EM (2005) Genomic sequencing of Pleistocene cave bears. Science 309: 597–599

3. Rohland N, Hofreiter M (2007) Comparison and optimization of ancient DNA extraction. Biotechniques 42:343–352
4. Rohland N, Siedel H, Hofreiter M (2010) A rapid column-based ancient DNA extraction method for increased sample throughput. Mol Ecol Resour 10:677–683
5. Rohland N, Hofreiter M (2007) Ancient DNA extraction from bones and teeth. Nat Protoc 2:1756–1762
6. Loreille OM, Diegoli TM, Irwin JA, Coble MD, Parson TJ (2007) High efficiency DNA extraction from bone by total demineralization. Forensic Sci Int Genet 1:191–195

Chapter 4

Case Study: Recovery of Ancient Nuclear DNA from Toe Pads of the Extinct Passenger Pigeon *

Tara L. Fulton, Stephen M. Wagner, and Beth Shapiro

Abstract

A variety of DNA extraction methods have been employed successfully to extract DNA from museum specimens. Toe pads are a common source of ancient DNA in birds, as they are generally not an informative character and can be removed without significant destruction of precious specimens. However, the DNA in these tissues is often highly degraded, both by natural postmortem decay and due to treatment by preservatives. In this case study chapter, we describe the use of both a commercial DNA extraction method and a silica-based method to extract ancient DNA from desiccated toe pads from the extinct passenger pigeon, *Ectopistes migratorius*. Successful amplification of nuclear DNA was achieved from both methods, representing the first nuclear DNA sequence recovered from this extinct species. We describe simple modifications to both protocols that we employed during the DNA extraction process.

Key words: Columbidae, Pigeons, Toe pads, DTT, Dithiothreitol, Ancient DNA, Silica extraction

1. Introduction

Specimens from museums and natural history collections are an important source of genetic information from the recent and distant past (2). Subsamples taken from museum-stored bones, skins, and other remains for ancient DNA analysis are precious, and therefore, the methods used to extract DNA from these should be efficient both in limiting the destructiveness of the subsampling and in the amount of DNA recovered. For museum-preserved

---

* *Note*: In the case study presented in this chapter, we describe the extraction of DNA from toe pads of museum-preserved specimens of the passenger pigeon, *Ectopistes migratorius*, using a method similar to that presented in Chap. 3. Other methods, such as those described in Chap. 2, may also be appropriate to extract DNA from this type of sample. We discuss specific challenges associated with applying this extraction method to ancient toe pad samples, including the use of dithiothreitol (DTT) for tissue dissolution. For more information, see the original publication of the scientific results in (1).

Beth Shapiro and Michael Hofreiter (eds.), *Ancient DNA: Methods and Protocols*, Methods in Molecular Biology, vol. 840,
DOI 10.1007/978-1-61779-516-9_4, © Springer Science+Business Media, LLC 2012

birds, a common source of DNA is toe pads. Toe pads are thick layers of tissue that can be easily removed with a sterile scalpel blade (3). Only a single toe pad is necessary for analysis, so that while toe pads are not generally phylogenetically informative, the toe pad from the opposing foot remains intact for future analysis. Problematically, museum specimens are often subjected to harsh chemicals as part of the process of preservation and, although toe pads may receive less focused treatment (3), these chemicals may complicate DNA extraction and downstream molecular biology reactions.

Passenger pigeons (*Ectopistes migratorius*) were once the most abundant birds in North America, comprising an estimated 20–40% of the total avian population and flock census estimates in the billions (4). However, European settlement of the region led to the rapid decline of the species over the course of the seventeenth to nineteenth centuries, with the death of the last known individual in captivity in 1914. Overhunting and habitat loss through deforestation are believed to be responsible for the extinction of the passenger pigeon, which occurred despite the enactment of some of the earliest conservation legislation in the US, which unfortunately was widely ignored (4).

Despite widespread public attention, very little is known about the evolutionary history of the passenger pigeon. Morphological analyses place it within the radiation of the New World mourning doves (*Zenaida*) (5), while mitochondrial DNA (mtDNA) suggests a relationship with the typical pigeons and doves (6). A recent reanalysis of the available mtDNA data (1,448 base pairs (bp) of the 12S rRNA and *cytochrome b* gene) with increased taxonomic coverage and an additional 169 bp of *ATP8* (also from the mitochondrion) suggested a close but not highly supported evolutionary relationship with the New World pigeons, *Patagioenas* (7). As mtDNA and nuclear DNA phylogenies are not always congruent, we targeted a nuclear intron to further examine passenger pigeon phylogeny. Analyses of these new data confirmed the sister relationship between the passenger pigeon and New World pigeons (*Patagioenas*) and provided moderately strong support for the *Ectopistes–Patagioenas* clade (1).

During the course of this study, we implemented two different extraction methods and designed a series of overlapping primers to obtain ancient nuclear intronic DNA. Here, we focus on comparison of the results of the two extraction methods, comment on technical hurdles encountered during the experiment, briefly discuss the strategy of designing PCR primers for targeting specific regions of ancient nuclear DNA, and suggest further improvements to the methods used based on subsequent analyses of this and other avian taxa.

2. Materials and Methods

We collected toe pad samples of *E. migratorius* from National Museums Liverpool (museum ID: Canon H.B. Tristram Collection LIV T17065). We performed DNA extraction and all pre-PCR work in a dedicated aDNA facility at the Pennsylvania State University, adhering to strict ancient DNA protocols at all stages. We cut each toe pad tissue into small pieces and isolated DNA using two different methods. First, we used the Qiagen DNeasy Tissue Kit (Qiagen), which we modified slightly to use 40 μL proteinase K and an extended initial lysis step of roughly 1 week, alternating between incubation at 50°C and at room temperature to ensure complete tissue lysis. We eluted the final extract into 200 μL and collected a second elution of 100 μL in case any DNA remained on the column membrane. This extraction represents the sample from which the full 445 bp of nuclear DNA was generated (1).

Second, we extracted a small section of toe pad using the silica protocol for DNA extraction from bones and teeth described in Chap. 3. Initially, we used in 10 mL of 0.5 M EDTA and proteinase K as per the protocol, incubating the sample at room temperature with rotation overnight. After 24 h, the toe pad remained largely intact, so we added an additional 5 mg of proteinase K and incubated the sample at room temperature with rotation for a further 24 h. As the toe pad still had not dissolved after this step, we then moved the solution to an incubator set at 50°C with rotation for a third 24 h period, then changed the temperature to 30°C for a fourth day of incubation with rotation. We added another 10 mg of proteinase K and continued to incubate the sample at 30°C with rotation for 4 more days. Finally, we added an additional 10 mg of proteinase K and incubated the sample with rotation for a final 24 h at 50°C, which eventually yielded a completely dissolved toe pad. We then processed the final solution following the silica extraction protocol.

We selected intron 7 of the nuclear-encoded *fibrinogen beta chain (FGB)* gene for sequencing. Due to the level of fragmentation incurred by aDNA, we designed a series of overlapping primer sets (Table 1 in (1)) to amplify fragments no longer than 170 bp, including the primers. To design the primers, we downloaded and aligned all available *FGB intron 7* sequences from GenBank. We used SeqBuilder v8.1 (DNASTAR) to design primers, which we selected to bind in regions that are conserved across the *Columbidae*. When we could not identify conserved regions, primers were designed to match most closely previously sequenced species belonging to the typical pigeons and doves (6). We compared the results of the two extraction protocols by attempting to amplify nuclear DNA using primer set *FGB*-F6R7.

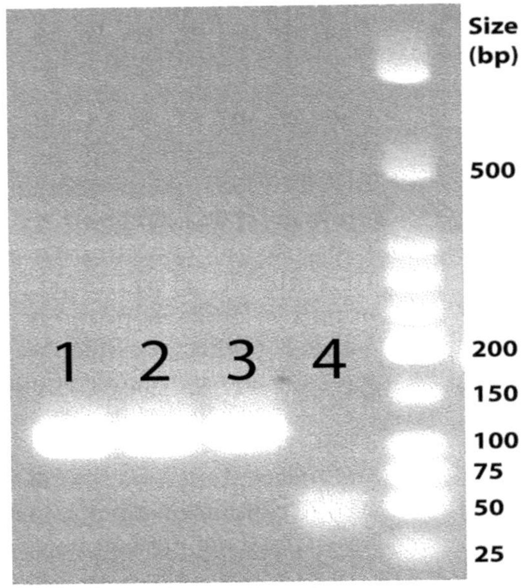

Fig. 1. Gel image of *FGB* F6R7 amplification. PCR products are visualized with ethidium bromide and UV illumination. Each *lane* contains 5 μL of 25 μL reactions with templates from (*left* to *right*): Qiagen first elution, Qiagen second elution, silica extraction, PCR negative, and 6 μL of a 50% dilution of NEB low molecular weight ladder. Primer–dimer can be observed in the PCR negative lane (*lane 4*), forming in the absence of template DNA.

We performed PCR amplifications in 25 μL reactions comprising 50 μg rabbit serum albumin, 0.25 mM dNTPs, 1× high fidelity buffer, 1.25 units Platinum *Taq* High Fidelity (Invitrogen), 3 mM $MgSO_4$, 1 μM of each primer, and 1 μL DNA extract. Cycling conditions were 94°C for 90 s, 60 cycles of 94°C for 45 s, 45 s at 48°C, 68°C for 90 s followed by 10 min of 68°C. We used a negative PCR reaction (containing no DNA extract) as the last reaction in each PCR. We amplified eluates from both extraction protocols, and results are shown in Fig. 1. Extraction negatives were previously run in separate reactions and were blank.

We performed maximum likelihood (ML) and ML bootstrap (MLBP) analyses using RAxML 7.0.4 (8) using the GTRGAMMA model. Bayesian phylogenies were estimated using MrBayes v3.2 (9, 10), applying the GTR+I+G model as selected using jModeltest v0.1.1 (11). Two runs of five million generations each were performed simultaneously, sampling every 200 generations and a 10% burnin. We assessed convergence by visualizing the traces in MrBayes and determining a potential scale reduction factor (PSRF) of ~1.00 for all parameters. Trees were visualized in FigTree v.1.3.1 (http://tree.bio.ed.ac.uk/software/figtree) and the ML tree with MLBP and Bayesian posterior clade probabilities (BPP) is shown in Fig. 2.

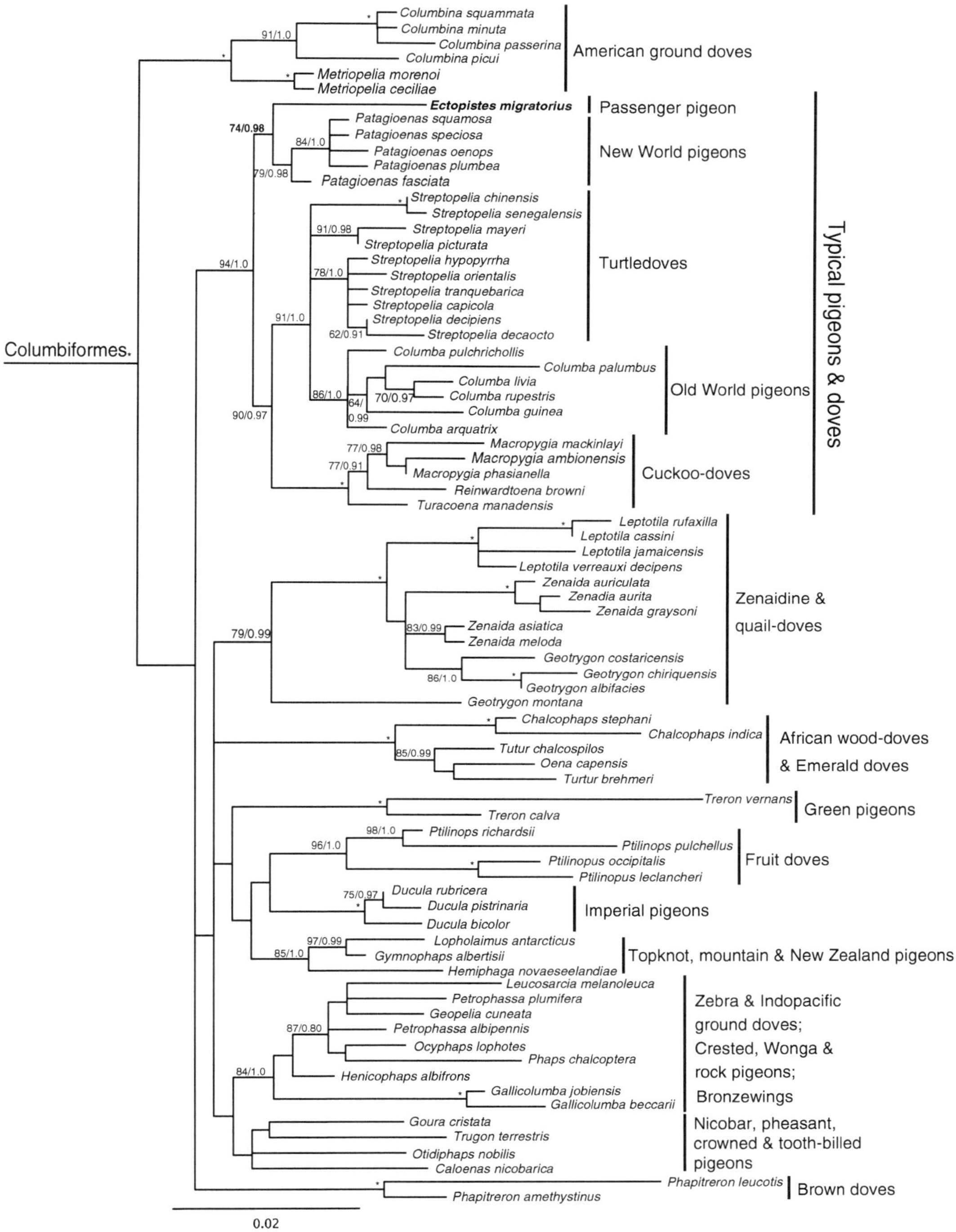

Fig. 2. Molecular phylogeny of *Columbidae* based on a maximum likelihood analysis of available sequences of *FGB intron 7*. Node support is indicated by maximum likelihood bootstrap (MLBP)/Bayesian posterior probability (BPP). Nodes supported by 100% MLBP and BPP = 1.0 are indicated with a star; nodes with <75% MLBP and BPP <0.95 are unlabeled. Common names are indicated on the *right*, where applicable. Figure is adapted from (1).

3. Results and Discussion

Both the Qiagen kit and silica DNA extraction methods produced amplifiable nuclear DNA from pigeon toe pads (Fig. 1). Although a large elution volume was used in the extraction utilizing the Qiagen kit, a second elution from the same column produced a band of the same intensity as the other extractions (lane 2, Fig. 1), highlighting the utility of a second elution off the column either at the time of initial extraction or after cold, dry storage (this also works well for silica columns).

Despite the high success rate of DNA recovery and amplification from this sample, several steps could have been taken to increase DNA yield. In subsequent tests with toe pads from other specimens, prolonged exposure to heat, as was applied during the silica extraction, resulted in significant decreases in DNA yield compared to exposure to much shorter periods of warming (48 h). In these experiments, 20 μL of 1 M dithiothreitol (DTT) was added to the extraction buffer to aid in the dissolution of the tissue (B. Letts, personal communication). Following these results, our lab began to routinely include DTT in extractions of bird toe pads using the Qiagen DNeasy extraction kit, and we noted that this step significantly reduced the amount of time required to completely dissolve the tissue. When only proteinase K and heat (and EDTA, but this is included primarily for bone decalcification) are applied to dissolve tissue, this process can take over a week; with buffers including detergents to disrupt tissue (as in the Qiagen tissue lysis buffer), dissolution can still take several days. We now routinely perform toe pad extractions by rinsing the intact toe pad in 0.5 M EDTA to wash away inhibitors, and then follow with the Qiagen DNeasy tissue extraction protocol, modified slightly by (1) including 20 μL of 1 M DTT with the initial tissue lysis step, (2) very gently shaking the samples in a 50°C oven for 48 h, and (3) adding an extra 20 μL of proteinase K solution after 24 h. We prefer warming in an incubator rather than a heat block, as the heat is more evenly distributed in the incubator. In our experience, most toe pad and desiccated tissue samples are dissolved after the first 24 h and all are dissolved after 48 h. The procedure then proceeds as per the manufacturer's protocol, although the elution volume is generally reduced to 50 μL, since the eluate can be diluted afterward. This procedure has yielded viable mtDNA (nuclear DNA not tested) from over 90% of pigeon toe pad specimens processed in our lab, including a specimen that was lacquered and several that were on display and exposed to light for many years.

The exceptional preservation of this specimen led to the recovery of a 445 bp sequence of *FGB intron 7*; the first nuclear sequence amplified from the extinct passenger pigeon (1). Maximum likelihood and Bayesian phylogenetic analyses indicate strong support for a sister relationship between the passenger pigeon and other New World pigeons (Fig. 2, adapted from (1)), confirming a single origin for pigeons in the New World. Phylogenetic analyses and sequence acquisition methods are described in detail in (1). In conclusion, we illustrate the possibility of obtaining nuclear DNA from well-preserved historic toe pad specimens from multiple DNA extraction methods to uncover information about an extinct species.

Acknowledgments

We thank Clemency Fisher and the National Museums Liverpool for providing *Ectopistes* material for analysis. Funding for this research was provided by the Pennsylvania State University to BS and an Undergraduate Discovery Grant to SMW.

References

1. Fulton TL, Wagner SW, Fisher C, Shapiro B (in press) Nuclear DNA from the extinct Passenger Pigeon (*Ectopistes migratorius*) confirms its phylogenetic placement within Columbinae. Ann Anat
2. Wandeler P, Hoeck PEA, Keller LF (2007) Back to the future: museum specimens in population genetics. Trends Ecol Evol 22:634–642
3. Mundy NI, Unitt P, Woodruff DS (1997) Skin from feet of museum specimens as a non-destructive source of DNA for avian genotyping. Auk 114:126–129
4. Schorger AW (1973) The passenger pigeon; its natural history and extinction. University of Oklahoma Press, Norman
5. Goodwin D (1967) Pigeons and doves of the world. British Museum (Natural History), London
6. Shapiro B, Sibthorpe D, Rambaut A, Austin J, Wragg GM, Bininda-Emonds ORP, Lee PLM, Cooper A (2002) Flight of the dodo. Science 295:1683
7. Johnson KP, Clayton DH, Dumbacher JP, Fleischer RC (2010) The flight of the passenger pigeon: phylogenetics and biogeographic history of an extinct species. Mol Phylogenet Evol 57:455–458
8. Stamatakis A (2006) RAxML-VI-HPC: maximum likelihood-based phylogenetic analyses with thousands of taxa and mixed models. Bioinformatics 22:2688–2690
9. Huelsenbeck JP, Ronquist F (2001) MRBAYES: Bayesian inference of phylogenetic trees. Bioinformatics 17:754–755
10. Ronquist F, Huelsenbeck JP (2003) MrBayes3: Bayesian phylogenetic inference under mixed models. Bioinformatics 19:1572–1574
11. Posada D (2008) jModelTest: Phylogenetic model averaging. Mol Biol Evol 25:1253–1256

Chapter 5

Extraction of DNA from Paleofeces

Melanie Kuch and Hendrik Poinar

Abstract

Paleofeces are the nonmineralized remains of dung from extant and extinct fauna. They represent a surprisingly large proportion of fossil remains recovered from cave sites across the world. Paleofeces contain the DNA of the defecator as well as the DNA of ingested plant and animal remains. To successfully extract DNA from paleofeces, a balance must be achieved between the minimization of DNA loss during extraction and the removal of coeluates that would otherwise inhibit the *Taq* DNA polymerase during downstream applications. Here we present a simplified version of a protocol to extract DNA from paleofecal remains.

Key words: DNA extraction, Feces, Ancient DNA, Silica DNA-purification

1. Introduction

Paleofeces, or as they are sometimes labeled, coprolites, are the nonmineralized remains of feces. As paleofeces are nonhardened fossils, they are typically assumed to be rare in the fossil record. However, deposits of megafaunal dung, namely that of the extinct ground sloth *Nothrotheriops shastensis* found in caves of the American southwest, rival in extent the vast deposits of mammoth bone and teeth in the permafrost (1).

Paleofeces are most often found within caves and rock shelters (2), although some have been found at open-air sites. Paleofeces likely make up a large percentage of the sediment found within cave floors, and to some degree, that found within permafrost soils as well. This may explain the success in retrieving the DNA of past inhabitants of caves (3) and megafauna of the high arctic (4).

The first successful extraction of DNA from paleofeces involved the use of PTB, a thiazolium salt shown to be successful in the reversal of Maillard-induced cross-links from reducing sugars (5–8). While it is unclear to what degree the PTB unlinks these nucleic

Beth Shapiro and Michael Hofreiter (eds.), *Ancient DNA: Methods and Protocols*, Methods in Molecular Biology, vol. 840, DOI 10.1007/978-1-61779-516-9_5, © Springer Science+Business Media, LLC 2012

acid/protein complexes, we have noted distinct improvements in total DNA yields when including PTB in ancient DNA extractions. Since the first publication reporting the successful extraction of DNA from paleofeces, subsequent reports have used these techniques to isolate not only the DNA of the defecator, but also DNA from the ingested contents (9, 10); please see ref. (11) for a more comprehensive review.

Below, we present the most up-to-date extraction protocol for the recovery of DNA from paleofecal remains. Since the original publication (12), we have assessed the relative success of various modifications using quantitative PCR assays and by measuring inhibition as gauged by the delaying of quantification cycles (13). The protocol can be optimized for the content of the dung (plant material versus a more meat-based diet). As with many fossil and subfossil samples, feces contain potent inhibitors with similar molecular weights, sizes and charges to DNA. The extraction of DNA from feces therefore requires achieving a balance between minimizing DNA loss during extraction and removing coeluates that would otherwise inhibit downstream applications. We use a chaotropic salt/silica-based procedure that is well known for its ability to remove inhibitors (14).

2. Materials

2.1. Laboratory Supplies (per Sample)

Forceps	1×
Small weight boat	1×
Scalpél blade	1×
15-mL tube	3×
1.5-mL tube	2×
0.5-mL tube	1×
pH paper	1+

2.2. Laboratory Equipment

1. Pipettes 1–1,000 μL.
2. Vortex.
3. Incubator with rotating wheel, or rotary mixer or similar device capable of being placed in an incubator.
4. Microcentrifuge as well as a large swing bucket centrifuge for 15 mL.
5. Shaking/heating block.

2.3. Solutions and Buffers (per Sample)

1. GuSCN extraction-buffer (14 mL per sample): 6-M guanidine thiocyanate. (GuSCN); 20 mM Tris-HCl, pH 8.0; 0.5% sodium lauroyl sarcosinate (Sarcosyl); 8 mM Dithiothreitol (DTT); 4% Polyvinylpyrrolidone (PVP); 10 mM *N*-phenacyl thiazolium bromide (PTB).
2. L6-buffer (4 mL per sample): 5 M GuSCN; 0.05 M Tris-HCl, pH 8.0; 0.0225 M Natrium chloride (NaCl); 0.02 M Ethylenediaminetetraacetic acid (EDTA), pH 8.0; 1.25% Triton X-100; (add 200 μL Silica, vortex, let sit).
3. Silica solution (50 μL per sample, see Subheading 3.1).
4. L2-buffer (1–2 mL per sample): 5 M GuSCN; 0.05 M Tris-HCl, pH 8.0; 0.0225 M NaCl (add 200 μL Silica, vortex, let sit).
5. New Wash substitute (1–2 mL per sample): make an 80% ethanol solution using 1× TE (10 mM Tris-HCl and 1 mM EDTA (pH 7.5)).
6. 0.1× TE pH 8.0 plus 0.05% Tween-20 (60 μL per sample).
7. Glacial Acetic acid (~15 μL per sample).

3. Methods

3.1. Preparing the Silica Suspension

1. Weigh out 4.8 g silica and place it into a 50-mL gamma-sterilized tube.
2. Add double-distilled (dd) H_2O to the tube containing the silica to 40 mL, vortex for 2 min, then let sit for 24 h at room temperature in the dark.
3. Carefully remove 35 ml of supernatant (without distrurbing the pellet) and discard.
4. Add ddH_2O to 40 mL, vortex for 2 min, let sit for 6 h at room temperature in the dark.
5. Carefully remove 36 ml of supernatant (without disturbing the pellet) and discard.
6. Add 48 μL 30% hydrochloric acid (keep pH acidic <3).
7. Resuspend the pellet, and aliquot approx. 200 μL of solution into separate tubes for later use.
8. Store in the dark at +4°C.
9. Prior to use, vortex to resuspend any pelleted material.

3.2. Paleofeces DNA Extraction

1. Add approximately 1 g of fecal material to a small weighing boat (see Note 1).
2. Cut the fecal remains into small pieces using scalpel blades (see Note 2).

3. Add fecal material to a final volume of 14 mL of the GuSCN extraction-buffer in a 15-mL tube and incubate, rotating overnight at 37°C in the dark (see Notes 1, 3 and 4).
4. Centrifuge at maximum speed for 5 min and transfer supernatant to a clean 15-mL tube.
5. Centrifuge again at maximum speed for 5 min and transfer supernatant to 4 mL of room temperature *pre-incubated* L6-buffer and 50 μL of Silica for at least 15 min (see Notes 5–8).
6. Adjust the pH to ~5 by adding glacial acetic acid (see Note 9).
7. Incubate while rotating at room temperature for 3 h in the dark (see Note 10).
8. Centrifuge for 5 min at maximum speed and discard the supernatant.
9. Add 1 mL of L2-buffer, resuspend, and transfer solution to a 1.5-mL microcentrifuge tube (see Note 11).
10. Centrifuge for 30 s and discard the supernatant.
11. If the solutions are still heavily colored, repeat steps 9–10.
12. Add 1 mL of wash buffer and resuspend the silica.
13. Centrifuge for 30 s and discard supernatant.
14. Centrifuge again briefly and remove all remaining liquid with a pipette.
15. Dry the remaining silica pellet in a heating block for 5 min at 56°C (to remove residual ethanol).
16. Add 60 μL of 0.1× TE pH 8.0 plus 0.05% Tween-20 and incubate with agitation for 15 min at 56°C.
17. Centrifuge for 3 min at maximum speed and transfer supernatant to a 0.5-mL tube.
18. Centrifuge again for 3 min at maximum speed and transfer supernatant to a clean 1.5-mL tube (see Note 12).
19. Freeze at −20°C.
20. Although counterintuitive, thaw the extract after completely freezing and make 10 μL aliquots and refreeze and store at −80 or −20°C (see Note 13).
21. Prepare a 1:10 dilution from the extract for PCR.

4. Notes

1. If less than 1 g of material is available, it is possible to scale the entire procedure down to 100 or 50 mg of sample using 1.75 mL of the GuSCN extraction-buffer. Volumes for L6-buffer + silica and all wash buffers remain the same.

2. Cut the feces into smaller pieces to allow for more coverage of the surface area of the feces with the extraction-buffer.
3. Add the PTB directly to the extraction buffer just before using it.
4. While it may seem odd, it appears (although the evidence for this is not statistically significant) to be better to add the fecal remains to the solution rather than add the solution to the fecal remains. We suspect this has something to do with the molecular availability of embedded nucleic acids and their accessibility to the salts in the buffer.
5. It appears to be beneficial to avoid the transfer of any solid material into the binding buffer (L6).
6. We have noticed a significant improvement in DNA recovery when silica and L6-buffer are incubated together prior to adding the supernatant from the GuSCN extraction-buffer/sample mix. Pre-incubate on a rotating wheel to insure proper mixing of silica and buffer.
7. We have attempted various compositions of extraction-buffer, for example, phosphate-based buffers and other chaotropic salts such as sodium periodate; none appear to be as successful in achieving the balance between DNA recovery and inhibitor removal as the buffer presented here.
8. Make up exactly enough fresh buffer for each use, and do not store the buffer, as guanidinium is light and temperature-sensitive and thus loses efficacy as it ages.
9. Keeping the buffers acidic is important (15, 16). Measure the pH of the solution after the sample has been added and adjust accordingly. The guanidinium must remain protonated to be an effective bridge between the phosphates on the DNA and the silica hydroxyl groups. In addition, alkaline conditions are far more conducive to DNA backbone degradation.
10. In our experience, extending the incubation time has not significantly altered total DNA yield.
11. It is wise to resuspend the silica pellet by placing the tip of the pipette at the edge of the pellet in the base of the tube and pipetting up and down slowly. Be careful not to allow the solution to bleed over the edge of the tube. While 50 μL of silica is sufficient for more than 10 μg of DNA, the silica clearly gets "clogged" with other polar molecules. Using less than 50 μL is therefore not advised. However, in our experience, increasing the amount of silica above 50 μL has not been shown to yield quantitatively more DNA.
12. Any silica remaining in the DNA solution may inhibit downstream reactions. As a precaution, spin down the extract before taking an aliquot for subsequent PCR. See ref. (17) for other helpful tips.

13. *Storage*: It is wise to freeze and thaw the extract post purification and prior to PCR. It appears that many of the inhibitors are precipitated out of solution during this step.

References

1. Martin P (1975) Sloth droppings. Nat Hist 74–78
2. Sobolik K (2003) Archaeobiology. AltaMira, Walnut Creek
3. Hofreiter M, Mead JI, Martin P, Poinar HN (2003) Molecular caving. Curr Biol 13:R693–R695
4. Willerslev E, Hansen AJ, Binladen J, Brand TB, Gilbert MT, Shapiro B, Bunce M, Wiuf C, Gilichinsky DA, Cooper A (2003) Diverse plant and animal genetic records from Holocene and Pleistocene sediments. Science 300:791–795
5. Cerami A, Vlassara H, Brownlee M (1987) Glucose and aging. Sci Am 256:90–96
6. Vasan S, Zhang X, Zhang X, Kapurniotu A, Bernhagen J, Teichberg S, Basgen J, Wagle D, Shih D, Terlecky I, Bucala R, Cerami A, Egan J, Ulrich P (1996) An agent cleaving glucose-derived protein crosslinks in vitro and in vivo. Nature 382:275–278
7. Wolffenbuttel BH, Boulanger CM, Crijns FR, Huijberts MS, Poitevin P, Swennen GN, Vasan S, Egan JJ, Ulrich P, Cerami A, Levy BI (1998) Breakers of advanced glycation end products restore large artery properties in experimental diabetes. Proc Natl Acad Sci U S A 95:4630–4634
8. Poinar GO Jr (1998) Trace fossils in amber: a new dimension for the ichnologist. Ichnos 6:47–52
9. Hofreiter M, Poinar HN, Spaulding WG, Bauer K, Martin PS, Possnert G, Pääbo S (2000) A molecular analysis of ground sloth diet through the last glaciation. Mol Ecol 9:1975–1984
10. Poinar HN, Kuch M, Sobolik KD, Barnes I, Stankiewicz AB, Kuder T, Spaulding WG, Bryant VM, Cooper A, Pääbo S (2001) A molecular analysis of dietary diversity for three archaic Native Americans. Proc Natl Acad Sci U S A 98:4317–4322
11. Pääbo S, Poinar H, Serre D, Jaenicke-Despres V, Hebler J, Rohland N, Kuch M, Krause J, Vigilant L, Hofreiter M (2004) Genetic analyses from ancient DNA. Annu Rev Genet 38:645–679
12. Poinar HN, Hofreiter M, Spaulding WG, Martin PS, Stankiewicz BA, Bland H, Evershed RP, Possnert G, Pääbo S (1998) Molecular coproscopy: dung and diet of the extinct ground sloth *Nothrotheriops shastensis*. Science 281:402–406
13. King C, Debruyne R, Kuch M, Schwarz C, Poinar H (2009) A quantitative approach to detect and overcome PCR inhibition in ancient DNA extracts. Biotechniques 47:941–949
14. Boom R, Sol CJ, Salimans MM, Jansen CL, Wertheim-van Dillen PM, van der Noordaa J (1990) Rapid and simple method for purification of nucleic acids. J Clin Microbiol 28:495–503
15. Fujiwara M, Yamamoto F, Okamoto K, Shiokawa K, Nomura R (2005) Adsorption of duplex DNA on mesoporous silicas: possibility of inclusion of DNA into their mesopores. Anal Chem 77:8138–8145
16. Melzak K, Sherwood C, Turner R, Haynes C (1996) Driving forces for DNA adsorption to silica in perchlorate solutions. J Colloid Interface Sci 181:635–644
17. Rohland N, Hofreiter M (2007) Comparison and optimization of ancient DNA extraction. Biotechniques 42:343–352

Chapter 6

DNA Extraction from Keratin and Chitin

Paula F. Campos and Thomas M.P. Gilbert

Abstract

DNA extracted from keratinous and chitinous materials can be a useful source of genetic information. To effectively liberate the DNA from these materials, buffers containing relatively high levels of DTT, proteinase K, and detergent are recommended, followed by purification using either silica-column or organic methods.

Key words: Chitinous tissue, Keratinous tissue, Silica, DNA extraction,, Ancient DNA, Hair, Feathers, Nail, Cuticle, Exoskeleton

1. Introduction

The DNA preserved in ancient or historic keratinous and chitinous materials is generally of lower quality than that preserved in other tissues, due to extensive DNA degradation during tissue biogenesis. Despite this, keratinous (e.g., hair, nails, feather, hoof, and horn sheath) and chitinous (e.g., insect cuticles) tissues are becoming increasingly popular as sources of ancient DNA (aDNA) (e.g. (1–5)). Although not always available, these tissues offer advantages over bone or soft tissue in that their surfaces are relatively simple to decontaminate, and that they can often be sampled unobtrusively.

Key to the extraction of DNA from keratinous tissues is to break down the keratin in order to liberate the DNA. This is achieved by using digestion buffers that contain large amounts of detergents and reducing agents (e.g., SDS, DTT, or Cleland's reagent) and proteinase K. DNA is then purified from the solution using either a silica-based purification or by extraction with organic solvents followed by isopropanol purification.

Beth Shapiro and Michael Hofreiter (eds.), *Ancient DNA: Methods and Protocols*, Methods in Molecular Biology, vol. 840,
DOI 10.1007/978-1-61779-516-9_6, © Springer Science+Business Media, LLC 2012

2. Materials

Prepare the digestion buffer using molecular biology reagents at room temperature, using appropriate anti-contamination controls (e.g., filter-tipped pipettes, DNA free consumables, etc.). To minimise the risk of contamination, we recommend purchasing ready-made stock solutions as opposed to making stock solutions in the lab.

2.1. General Requirements

1. Bleach solution, diluted in H_2O to a final NaClO concentration of approximately 0.5%.
2. "Molecular Biology Grade" H_2O (ddH_2O).
3. Stable digestion buffer: 10 mM Tris buffer (pH 8.0), 10 mM NaCl, 5 mM $CaCl_2$, 2.5 mM EDTA (pH 8.0), 2% SDS. Store at 4°C.
4. 1 M Dithiothreitol (DTT) solution. Make up fresh for each digestion and discard unused solution (see Note 1).
5. Proteinase K solution (see Note 1).
6. Centrifuge(s) for 1.5/2 mL (>10,000 × *g*) and 15-mL tubes (>3,000 × *g*), dependent on size of digestions and purifications to be performed.
7. Oven prepared at 55°C, within which a rotary device (below) can be placed.
8. Rotary mixer, wheel or similar device to keep samples constantly in motion during incubation steps.
9. Tabletop vortex.
10. Sterile 1.5-mL and 15-mL tubes, depending on the size of the extraction being performed.

2.2. For Silica-Column Purification (3.3)

1. Qiaquick DNA purification kit (Qiagen, Valencia, CA) including "Qiaquick" silica columns, and buffers "PB", "PE", and "EB."

2.3. For Organic Purification (3.4)

1. Tris-buffered phenol, pH 8.0.
2. Chloroform.
3. Isopropanol.
4. 3 M sodium acetate, approximate pH 5.
5. (*Optional*) DNA precipitation "carrier," e.g., Glycoblue (Ambion, Inc., Austin, TX).
6. "Molecular Biology Grade" ethanol, 85%.
7. TE elution buffer: 10 mM Tris–HCl, 1 mM EDTA (pH 8.0).

3. Methods

Carry out all procedures at room temperature unless otherwise specified. Always incorporate extraction blanks in the analysis. This protocol assumes the use of pure keratin or chitin, e.g. hair, horn, nail, feather, arthropod exoskeleton/wing carapaces.

3.1. Tissue Pre-Preparation

1. For most materials, proceed directly to step 2. For large pieces of nail or horn, drill a suitable amount of powder (e.g., 100 mg) directly from the specimen and collect the powder in an appropriate container.
2. *For non-powdered material*, clean the tissue via a brief wash in dilute bleach solution, taking care to remove all obvious sources of contaminant matter. *For powdered material*, clean the powder by immersing it in the bleach solution for 10–20 s, then briefly centrifuge the mixture to pellet the powder. Pour off the bleach.
3. Rinse material several times in ddH_2O to remove all traces of bleach (see Note 2). *For powdered material*, use a vortex to ensure the pellet from step 2 is homogenised after adding the water. After 10–20 s of incubation, re-pellet the powder. Pour off the ddH_2O then repeat.

3.2. Tissue Digestion

1. Add 40 μL 1 M DTT solution and 100 μL proteinase K solution per 860 μL stable digestion buffer to make the active digestion buffer. Mix well (see Note 3).
2. Add a suitable amount of digestion buffer to the sample (see Note 3). Vortex briefly to ensure that any pelleted powder is homogenised in the solution. Incubate the sample plus buffer overnight at 55°C with gentle agitation.
3. Keratinous samples may not fully digest after this incubation. If full digestion is required, add an additional 40 μL 1 M DTT solution and 100 μL proteinase K solution to the mixture, vortex briefly, and return to incubation with agitation for at least 1 more hour. Chitinous samples rarely fully digest; however, in both tissues, DNA is usually liberated into solution even if digestion does not appear to be complete upon visual inspection.
4. Proceed to DNA purification using either the silica (see Subheading 3.3) or organic (see Subheading 3.4) method (see Note 4).

3.3. DNA Purification: Silica Method (see Note 5)

1. Centrifuge the digestion mixture for 3–5 min at high speed (>10,000 × *g*) to pellet any solid remains. Carefully pipette the liquid fraction of the digestion into a new tube. Avoid transferring any solids that may block the spin filter.

2. Add 5 volumes Qiaquick buffer PB to the solution.
3. Mix thoroughly.
4. Add 700 μL of this mixture to the Qiaquick spin column.
5. Centrifuge for 1 min at 6,000 × *g*. This speed is useful to limit how much target DNA passes through the filter without binding. However, if the liquid does not pass through the filter in 1 min, the speed can be increased.
6. Empty the liquid waste from the spin column (see Note 6). Repeat steps 5–6 with the remaining PB buffer-digestion mix until all the liquid passes through the spin column.
7. Add 500 μL Qiagen wash buffer PE to the filter.
8. Centrifuge for 1 min at 10,000 × *g*. Empty the waste and repeat if extra purity is required.
9. Centrifuge for 3 min at maximum speed to dry the filter. Any residual ethanol from the PE buffer will inhibit downstream applications.
10. Place the filter in a new 1.5-mL tube. Add 50–100 μL Qiagen elution buffer EB directly to the centre of the filter, and leave at room temperature for 5 min (see Note 7). EB can be replaced with molecular biology grade ddH_2O (pH 7–8).
11. Centrifuge for 1 min at maximum speed to collect the EB and DNA.

3.4. DNA Purification: Organic Extraction (see Note 8)

1. Add phenol (pH 8) to the digestion mix at a ratio of 1:1 with the total digestion volume.
2. Agitate gently at room temperature for 5 min.
3. Centrifuge for 5 min to separate the layers. The speed of centrifugation will depend on the volume of the digestion mix, the centrifuge capacity, and the maximum speed designation of the tubes being used. It is generally advisable to use the maximum speed possible. If after 5 min the layers have not fully separated, extend the centrifugation time.
4. Carefully remove the upper aqueous layer. Be careful not to remove the protein-containing interface. Discard the lower, phenol layer (see Note 8).
5. Add to 1 volume of new phenol. Repeat steps 2–4 in Subheading 3.4. After the second centrifugation, add the aqueous layer to 1 volume chloroform.
6. Agitate gently at room temperature for 5 min.
7. Centrifuge for 5 min to separate the layers. Remove the upper aqueous layer. Discard the lower, chloroform layer (see Note 8).
8. Add 0.6–1 volume isopropanol and 0.1 volume 3 M sodium acetate (approx. pH 5). A small amount of commercial carrier

solutions can also be added if required to facilitate pellet visualisation, such as Glycoblue (Ambion, Inc., Austin, TX), following the manufacturers' guidelines. Mix well (see Note 9).

9. Immediately centrifuge at high speed (>10,000 × *g*) for 30 min at room temperature.
10. Immediately following centrifugation, decant the liquid from the tube carefully. The DNA will have precipitated into a pellet at the bottom of the tube and may not be visible.
11. To rinse the pellet, gently add 500–1,000 μL 85% ethanol, slowly invert the tube once, then centrifuge for 5 min at high speed.
12. Gently decant the ethanol. Repeat if necessary.
13. All ethanol must be removed from the pellet as any residual ethanol will inhibit downstream applications. This can be achieved by using a small bore pipette and by briefly incubating the dry pellet at a relatively high temperature (e.g., 55–75°C).
14. Resuspend the pellet in elution TE buffer or ddH_2O. If the pellet has become very dry, this may require leaving the pellet at room temperature in the liquid for 5–10 min, followed by gentle pipetting (see Note 10).

4. Notes

1. Neither DTT nor proteinase K are stable once added to the active digestion solution, thus the active buffer needs to be freshly made for each digestion. At 4°C, the SDS will precipitate out of solution. Prior to the addition of DTT and proteinase K, the buffer should be warmed up until the SDS is fully dissolved.
2. Any bleach carryover will degrade the DNA and reagents in subsequent steps of the DNA extraction, thus it is extremely important that bleach is removed completely.
3. The volume of digestion buffer needed is sample dependent, but generally should be at least sufficient to cover the surface of the material.
4. DNA can be purified from the digestion mixture in a number of different ways. Selecting a method depends ultimately on convenience and user preference. For small volumes, silica spin-columns are convenient, but for larger volumes these rapidly become very labour-intensive. For larger volumes of digestion mix (e.g., >1 mL), organic extractions are often preferable, in particular if large amounts of undigested melanin, dirt or

other material are present in solution, as these tend to block silica filters. For a silica protocol, refer to Subheading 3.3. For organic purification, refer to Subheading 3.4.

5. As recommended by Yang et al. (6), Qiagen's "Qiaquick" PCR clean-up kits are an excellent and quick tool for purifying DNA. The instructions in the kit manual can be followed almost directly if one replaces the phrase "PCR product" with "DNA extract". The only change we recommend is the modification of the centrifugal speeds.
6. Qiagen buffers contain guanidinium salts, and relevant local disposal regulations should be consulted.
7. The volume of EB to use in this step depends on final concentration of DNA required and can be modified.
8. Organic extractions use phenol and chloroform to help purify the DNA. Both phenol and chloroform are toxic, and phenol in particular is extremely dangerous. Neither should be used without appropriate training. Always handle both liquids and their containers with extreme care, using appropriate face, hand, and body protection. Do not handle using latex gloves, as these are permeable to phenol and chloroform; use only nitryl gloves. The fumes of both chemicals are dangerous; therefore, these steps should always be performed in a vented fume hood. Disposal of both requires conformation to specific regulations, thus relevant local disposal regulations should be consulted.
9. Isopropanol precipitation is most effective at relatively high centrifugal forces and in small tubes (the DNA pellet is easiest to see and resuspend if 1.5-mL tubes or smaller are used). If large volumes are to be precipitated, we recommend first concentrating the liquid, for example with a centrifugal concentrator such as an Amicon centricon (Millipore, Billerica, MA) with 30 kD or less molecular weight cut-off.
10. Melanin pigments often copurify with the DNA and coprecipitate with the DNA during isopropanol precipitation. This results in a brown concentrated DNA pellet and a brown extract after resuspension. As melanin can inhibit enzymatic reactions (e.g., PCR), an additional purification step may be followed, for example using a silica procedure (e.g., see Subheading 3.3).

Acknowledgments

MTPG was supported by the Danish National Science Foundation's "Skou" grant program.

References

1. Bonnichsen R, Hodges L, Ream W et al (2001) Methods for the study of ancient hair: radiocarbon dates and gene sequences from individual hairs. J Archaeol Sci 28:775–785
2. Gilbert M, Wilson A, Bunce M et al (2004) Ancient mitochondrial DNA from hair. Curr Biol 14:463
3. Rawlence N, Wood J, Armstrong K et al (2009) DNA content and distribution in ancient feathers and potential to reconstruct the plumage of extinct avian taxa. Proc Biol Sci 276:3395
4. Willerslev E, Gilbert MT, Binladen J et al (2009) Analysis of complete mitochondrial genomes from extinct and extant rhinoceroses reveals lack of phylogenetic resolution. BMC Evol Biol 9:95
5. King G, Gilbert M, Willerslev E et al (2009) Recovery of DNA from archaeological insect remains: first results, problems and potential. J Archaeol Sci 36:1179–1183
6. Yang DY, Eng B, Waye JS et al (1998) Technical note: improved DNA extraction from ancient bones using silica-based spin columns. Am J Phys Anthropol 105:539–543

Chapter 7

Case Study: Ancient Sloth DNA Recovered from Hairs Preserved in Paleofeces*

Andrew A. Clack, Ross D.E. MacPhee, and Hendrik N. Poinar

Abstract

Ancient hair, which has proved to be an excellent source of well-preserved ancient DNA, is often preserved in paleofeces. Here, we separate and wash hair shafts preserved in a paleofecal specimen believed to be from a Darwin's ground sloth, *Mylodon darwinii*. After extracting DNA from the recovered and cleaned hair using a protocol optimized for DNA extraction from keratinous substrates, we amplify 12S and 16S rDNA sequences from the DNA extract. As expected, the recovered sequences most closely match previously published sequences of *M. darwinii*. Our results demonstrate that hair preserved in paleofeces, even from temperate cave environments, is an effective source of ancient DNA.

Key words: Ancient DNA, Coprolite, Hair, Paleofaeces, Sloth

1. Introduction

Preserved hair is known to be an excellent source of ancient DNA (1–5). DNA extracted from paleofeces has been used in taxonomic analyses (6), dietary reconstructions (7–9), and to identify the presence of taxa in environments where the fossil record is incomplete (9, 10). Hair shafts in fecal samples (11, 12) may belong either to the defecator, to conspecific individuals as a result of grooming, or to prey. Ingested hair shafts can eventually be passed through the digestive tract, due to the durability of the keratinous exterior (1). Paleofeces may therefore represent an underutilized substrate for ancient hair that can be used for genetic research.

* *Note*: For the case study presented in this chapter, we describe DNA extraction and amplification from ancient hairs preserved in paleofeces using a method similar to that presented in Chap. 6.

Beth Shapiro and Michael Hofreiter (eds.), *Ancient DNA: Methods and Protocols*, Methods in Molecular Biology, vol. 840, DOI 10.1007/978-1-61779-516-9_7, © Springer Science+Business Media, LLC 2012

2. Materials and Methods

A paleofecal sample attributed to *Mylodon darwinii* that contained hairs within its matrix was collected in 2001 from Cueva del Milodón in southern Chile and stored thereafter at the American Museum of Natural History (New York, NY). The specimen is estimated to be ~13,000 years old, which is in line with other specimens from this locality (13). We extracted DNA from this specimen at the McMaster Ancient DNA Centre (Hamilton, ON, Canada), where modern and ancient processing facilities are spatially separated and sterile conditions are maintained to prevent contamination with modern DNA.

Using bleach and oven-sterilized tweezers, we removed six hair shafts of ~1–2 cm in length from the paleofecal sample. We washed the hairs in sterile H_2O three times to remove exterior debris. Using fresh scalpel blades, we cut the hairs into equal length pieces of ~0.5 cm and placed them in a clean 2-mL tube. We then implemented an extraction protocol similar to that presented in Chap. 6 (14). We added 1,200 μL of digestion buffer (1) to the tube containing the hair shafts, which we then incubated (along with a negative control) at 55°C. The hairs were fully digested after 10 h. We then added 500 μL of PCI (phenol/chloroform/isoamyl alcohol (25:24:1)) to the digested solution and shook the mixture gently for several minutes. We centrifuged the mixture at maximum speed for 5 min and transferred the aqueous phase to a new 2-mL tube. We repeated the process using 500 μL of chloroform to remove residual phenol. We then concentrated and washed the aqueous extract layer (~1,000 μL).

We primed Microcon filter cartridges (Millipore, Canada) with 100 μL of laboratory-made 0.1× TE buffer and added either the sample or the blank to specific cartridges (in sequential steps of 500 μL until the entire sample had been passed through the filter). We washed the filter membranes three times with 300 μL of 0.1× TE. Finally, we added 100 μL of 0.1× TE to each cartridge, placed it in a new collection tube, and agitated the cartridge for 5 min at 1,000 rpm on the heat block at room temperature. Finally, we inverted the Microcon cartridges and centrifuged at 1,000 × *g* for 3 min to collect the concentrated DNA. The final extraction and blank were frozen overnight, thawed, and vortexed for 20 s before use.

We performed PCR in 20 μL reaction volumes, using 3 μL of undiluted extract/blank. All reagents were thoroughly thawed and vortexed for 20 s before use. Primers were designed by eye using the *M. darwinii* sequence published by Höss et al. (13) (GenBank accession nos. Z48943 (12S) and Z48944 (16S)):

Md16SF 5′TAGGGATAACAGCGC-AATCC3′.

Md16SR 5′CGTAGGACTTTAATCGTTGA3′.

Md12S 5′CTGGGATTAGATA-CCCCACTAT3′.

Md12SR 5′GTCGATTATAGGACAGGTTCCTCTA3′.

With primers, the target fragments were 147 and 152 bp long for the 12S and 16S fragments, respectively. PCR conditions were as follows: initial denaturation at 95°C for 5 min, 40 cycles of denaturation for 30 s at 95°C, annealing for 30 s at 55°C, and extension at 72°C for 30 s, with a final extension at 72°C for 10 min. Both fragments were PCR amplified three times.

Following PCR, 4 μL of each amplification product were loaded onto a 2.5% agarose gel stained with ethidium bromide, along with 1 μL of 100 bp DNA ladder. The gel was run in an electrophoresis chamber and separated products were visualized under UV light. PCR products were cloned using a TOPO-TA cloning kit (Invitrogen, Canada). Insert-carrying colonies were identified and gently stabbed with a sterile 10-μL pipette tip to remove a small sample of bacteria. Each tip was soaked in 100 μL of sterile H_2O in PCR strip tubes, which were then heated at 95°C for 5 min on a thermocycler to lyse the bacteria. We then amplified the PCR inserts using M13 forward and reverse primers (provided in TOPO-TA cloning kits) and the previously described PCR and cycling profile, using 2 μL of the lysed bacteria mix as DNA template. PCR products were purified using AcroPrep 96 filter plates (Pall, USA), visualized on a gel, and approximately quantified using the DNA ladder.

We sequenced the purified and quantified PCR products using the M13 forward primer as per manufacturer's suggestions, with 1 μL of DNA, in 7 μL reactions, using 0.3 μL of BigDye ver1.1 (Applied Biosystems, Foster City, CA) and 1.5 μL buffer. We cleaned the cycle sequencing products and sent the DNA to the MOBIX sequencing facilities on McMaster University campus (Hamilton, ON, Canada) for sequencing. Finally, we visualized, aligned, and edited the resulting sequences and trace files using BioEdit (ver5.07) (15).

3. Results and Discussion

We obtained three independent PCR products each for fragments of the 12 and 16S rDNA genes. We cloned the products and sequenced 12 and 9 clones for 12S and 16S, respectively. From these, we derived consensus sequences, both of which match the sequence for *M. darwinii* from Höss et al. (13) (Figs. 1 and 2), but differ from other sloth taxa. Within the clones, we observed 24 C to T transitions and three G to A transitions. This type of sequence damage is likely the result of hydrolytic deamination (16) and is common in ancient specimens (16, 17). In this experiment, one

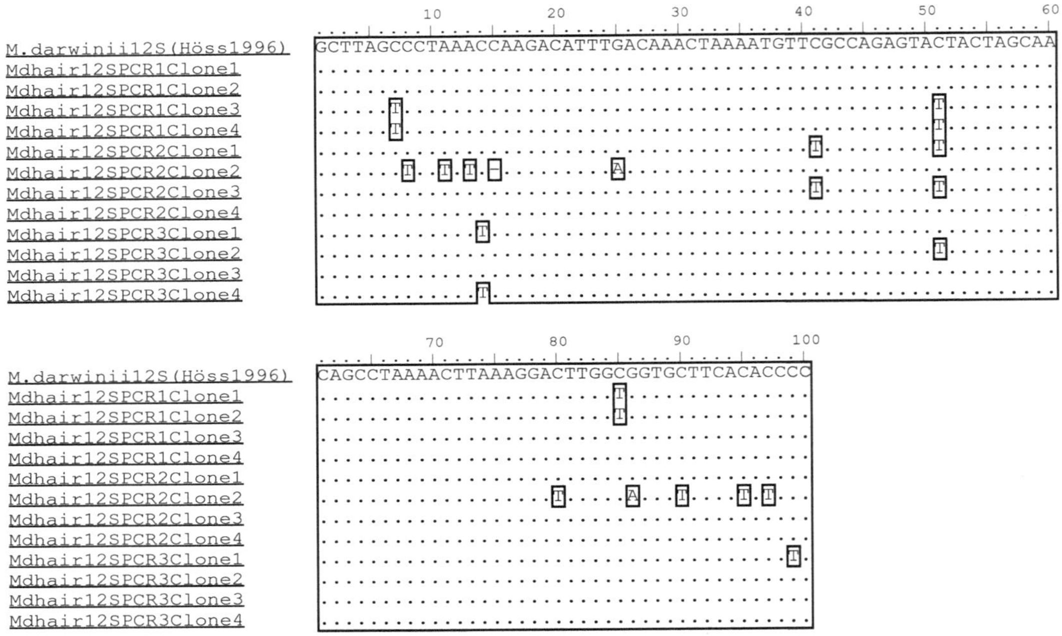

Fig. 1. Alignment of four cloned PCR products each from three different amplifications of 12S rDNA, originally amplified from a hair shaft belonging to *M. darwinii* that was isolated from a paleofecal sample.

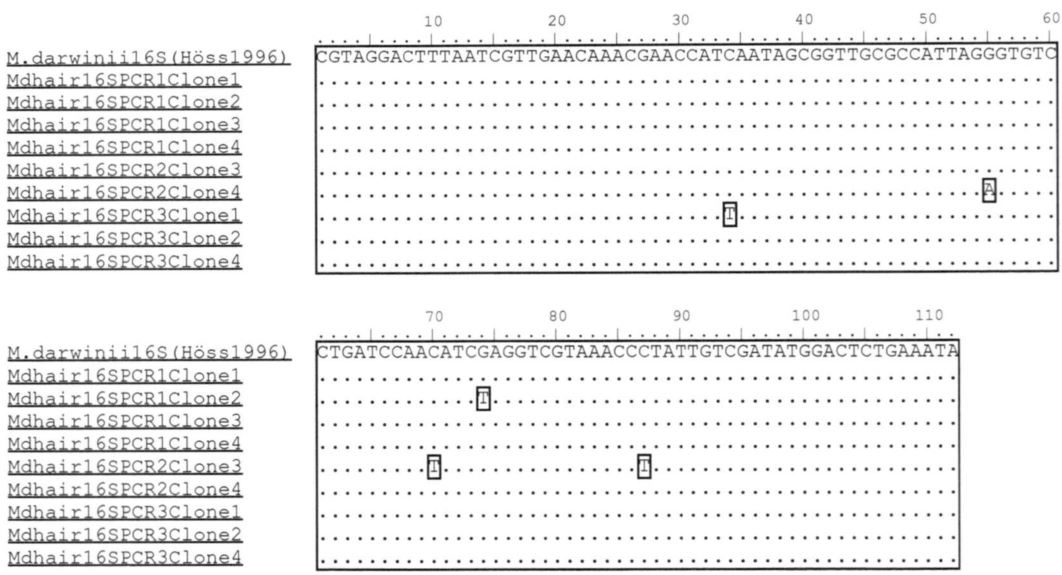

Fig. 2. Alignment of two to four cloned PCR products each from three different amplifications of 16S rDNA, originally amplified from a hair shaft belonging to *M. darwinii* that was isolated from a paleofecal sample.

clone sequence was particularly damaged: 12S PCR2 clone2 (Fig. 1) displays seven C to T transitions, two G to A transitions, and a deletion. Given that all damaged sites are the most common type of damage in ancient DNA, it is reasonable to assume that fragment is not an exogenous contaminant, but rather a highly degraded starting template, and perhaps also affected by jumping PCR.

The presence of *M. darwinii* DNA in hair shafts preserved within paleofeces reveals an additional source of ancient DNA for downstream analyses. Paleofeces are comprised of both a broad diversity of processed material and the defecator's own sloughed tissue (7–9). Separating the constituent materials prior to DNA extraction could facilitate downstream applications, such as targeted sequencing.

Hair shafts, if present in paleofeces, represent macroscopic packets of species-specific cells, potentially enriched with mtDNA (2) and relatively simple to separate, clean, and process. In addition, the gross structure of hair may significantly limit exogenous DNA contamination (18). Finally, the relatively simple process of separating and cleaning hair of fecal debris dramatically decreases the potential of coamplifying contaminating sequences from the paleofeces itself, including DNA from the defecator. This could add novel insights into, for example, the diets of carnivores (12), or conspecific oral grooming behaviors.

References

1. Gilbert MTP, Wilson AS, Bunce M, Hansen AJ, Willerslev E, Shapiro B, Higham TFG, Richards MP, O'Connell TC, Tobin DJ, Janaway RC, Cooper A (2004) Ancient mitochondrial DNA from hair. Curr Biol 14:R463–R464
2. Gilbert MTP et al (2008) Intraspecific phylogenetic analysis of Siberian woolly mammoths using complete mitochondrial genomes. Proc Natl Acad Sci U S A 105:8327–8332
3. Gilbert MTP et al (2008) Paleo-Eskimo mtDNA genome reveals matrilineal discontinuity in Greenland. Science 320:1787–1789
4. Miller W et al (2008) Sequencing the nuclear genome of the extinct woolly mammoth. Nature 456:387–390
5. Rasmussen M et al (2010) Ancient human genome sequence of an extinct Palaeo-Eskimo. Nature 463:757–762
6. Poinar HN, Kuch M, McDonald G, Martin P, Pääbo S (2003) Nuclear gene sequences from a Late Pleistocene sloth coprolite. Curr Biol 13:1150–1152
7. Poinar H, Hofreiter M, Spaulding G, Martin P, Stankiewicz A, Bland H, Evershed R, Possnert G, Pääbo S (1998) Molecular coproscopy: dung and diet of the extinct ground sloth *Nothrotheriops shastensis*. Science 281:402–406
8. Poinar HN, Küch M, Sobolik KD, Barnes I, Stankiewicz AB, Kuder T, Spaulding WG, Bryant VM, Cooper A, Pääbo S (2001) A molecular analysis of dietary diversity for three archaic Native Americans. Proc Natl Acad Sci U S A 98:4317–4322
9. Hofreiter M, Betancourt JL, de Sbriller AP, Markgraf V, McDonald HG (2003) Phylogeny, diet and habitat of an extinct ground sloth from Cuchillo Curá, Neuquén Province, southwest Argentina. Quat Res 59:364–378
10. Kuch M, Rohland N, Betancourt JL, Latorre C, Steppan S, Poinar HN (2002) Molecular analysis of an 11,700-year-old rodent midden from the Atacama Desert, Chile. Mol Ecol 11:913–924
11. Zhang W, Zhang Z, Xu X, Wei K, Wang X, Liang X, Zhang L, Shen F, Hou R, Yue B (2009) A new method for DNA extraction from FECES and hair shafts of the South China Tiger (*Panthera tigris amoyensis*). Zoo Biol 28:49–58

12. Backwell L, Pickering R, Brothwell D, Berger L, Witcomb M, Martill D, Penkman K, Wilson A (2009) Probable human hair found in a fossil hyaena coprolite from Gladysvale cave, South Africa. J Archaeol Sci 36:1269–1276
13. Höss M, Dilling A, Currant A, Pääbo S (1996) Molecular phylogeny of the extinct ground sloth *Mylodon darwinii*. Proc Natl Acad Sci U S A 93:181–185
14. Campos PF, Gilbert MTP (2011) DNA extraction from keratin and chitin. In: Shapiro B, Hofreiter M (eds) Ancient DNA. Springer, New York
15. Lindahl T (1993) Instability and decay of the primary structure of DNA. Nature 36:709–715
16. Hall TA (1999) BioEdit: a user-friendly biological sequence alignment editor and analysis program for Windows 95/98/NT. Nucleic Acids Symp Ser 41:95–98
17. Gilbert MTP et al (2007) Recharacterization of ancient DNA miscoding lesions: insights in the era of sequencing-by-synthesis. Nucleic Acids Res 35:1–10
18. Gilbert MTP, Menez L, Janaway RC, Tobin DJ, Cooper A, Wilson AS (2006) Resistance of degraded hair shafts to contaminant DNA. Forensic Sci Int 156:208–212

Chapter 8

Ancient DNA Extraction from Soils and Sediments

James Haile

Abstract

DNA contained in soils and sediments can provide novel insights into past environments and ecosystems. In this chapter, I describe an efficient and effective technique to extract total DNA from sediments in a manner that minimizes the coextraction of PCR-inhibitory compounds. I describe two different approaches: one that is suitable for large (up to 10 g wet weight) amounts of substrate, and a second that is more appropriate for small (up to 0.5 g) amounts of substrate. Finally, I discuss some of the obstacles that may be encountered in the process of extracting DNA from soils and sediments and suggest approaches to circumvent some common problems.

Key words: Sediment, Soils, Ancient DNA, Metagenomics, Environmental sampling, SedaDNA

1. Introduction

Sediments and paleosols have proven to be an excellent repository of ancient DNA of plants, fungi, and animals from both arctic and temperate biomes and from tropical and arid environments (1–5). However, the humic compounds and other organomineral complexes to which the extracellular DNA binds and which protect the DNA from extracellular, microbial DNases, and nucleases (6, 7) also inhibit PCR amplification. Therefore, any successful extraction of DNA from sediments will need to remove these substances.

Sediments are heterogeneous with respect to DNA distribution, and a compromise needs to be struck between using large volumes of sample so as to maximize the chance of recovering DNA and the resulting decrease in fine, temporal resolution that can occur when large samples are processed. Given this limitation, processing larger amounts of sediment tends to improve the success rate of extracting rare or low-copy number DNA. As large volume samples are not always available, I describe protocols for both large extractions and small extractions below.

Beth Shapiro and Michael Hofreiter (eds.), *Ancient DNA: Methods and Protocols*, Methods in Molecular Biology, vol. 840, DOI 10.1007/978-1-61779-516-9_8, © Springer Science+Business Media, LLC 2012

For large extractions (up to 10 g wet weight) of sediment, the PowerMaxSoil™ DNA Isolation Kit (Cambio) is recommended. In this protocol, up to 10 g wet weight of sediment sample is homogenized and cells lysed using a 50-mL tube containing garnet grits and extraction buffer. The lysate is then progressively cleaned of cellular debris by centrifugation and precipitation. Aqueous molecules coextracted with the DNA are then removed using silica spin columns.

For extractions of up to 0.5 g, a protocol that uses components from FastDNA® SPIN Kit for Soil for isolation (QBIOgene) is recommended. The soil sample is added to a 2-mL tube that contains glass beads. The tube is then shaken vigorously in the presence of an extraction buffer, to pulverize and lyse the samples. Lipids are removed using chloroform/octanol and the DNA-containing solution cleaned using silica spin columns.

2. Materials

Extraction of sedimentary ancient DNA (*sed*aDNA) should be carried out in a dedicated ancient DNA facility using established protocols (8).

2.1. Large Extraction

1. 100-μL, 1-mL, and 10-mL pipettes and tips.
2. 1.5-mL tubes (at least one per extract, depending upon final volume eluted).
3. 50-mL tubes (one per sample).
4. Rotary mixer, wheel, or similar device to keep samples constantly in motion during incubation steps, capable of holding 50-mL tubes.
5. Oven large enough to accommodate the rotary mixer.
6. Centrifuge capable of holding 50-mL tubes and reaching a force of 2,500 × *g*.
7. Vortex-Genie® Vortex and a Vortex Adapter capable of shaking two 50-mL tubes simultaneously (CamBio).
8. Garnet grit: aliquots provided in the PowerMax® Bead Tubes from the PowerMaxTM DNA Isolation Kit.
9. Bulat (9) extraction buffer: 0.02 g/mL Sarcosyl, 50 mM Tris–HCl (pH 8.0), 20 mM NaCl, 3.5% 2-mercaptoethanol, 50 mM 1,4-Dithio-L-threitol (DTT), 2 mM *N*-phenacylthiazone bromide (PTB), 0.8 g/mL Proteinase K (see Note 1).
10. Solutions C1–C6 from the PowerMaxSoil™ DNA Isolation Kit (Cambio).
11. HPLC grade water.

2.2. Small Extraction (Materials)

1. 100-mL and 1-mL pipettes and tips.
2. FastPrep® Instrument (Qbiogene).
3. FastPrep® Lysing Matrix E tubes.
4. 1.5-mL tubes (two per sample).
5. Microcentrifuge capable of reaching a force of 12,000 × *g*.
6. Rotary mixer, wheel, or similar device to keep samples constantly in motion during incubation steps, capable of holding 2-mL tubes.
7. Oven large enough to accommodate the rotary mixer.
8. Bulat (9) extraction buffer : 0.02 g/mL Sarcosyl, 50 mM Tris–HCl (pH 8.0), 20 mM NaCl, 3.5% 2-mercaptoethanol, 50 mM 1,4-Dithio-L-threitol (DTT), 2 mM *N*-phenacylthiazone bromide (PTB), 0.8 g/mL Proteinase K (see Note 1).
9. 5 M NaCl, Chloroform:Octanol (24:1).
10. PB-buffer (QIAGEN).
11. Salton wash 1 buffer (BIO 101).
12. Salton wash 2 buffer (BIO 101).
13. AW1 buffer (QIAGEN).
14. EB buffer (QIAGEN).
15. HPLC grade water.
16. Ice.

3. Methods

Sediments should be sampled in such a way so as to minimize the possibility of cross-contamination or contamination with modern DNA. Where possible, sediments should be frozen immediately after sampling.

3.1. Large Extraction

Carry out all procedures at room temperature unless otherwise stated.

1. Add 10 g wet weight of sediment to a 50-mL tube containing garnet grits.
2. Add 12 mL of Bulat extraction buffer or 15 mL of PowerMax® Bead Solution C2 (containing guanidine thiocyanate) (see Note 2).
3. Add 1.2 mL of C1 solution (sodium dodecyl sulfate solution) (see Note 3).

4. Vortex for 10 min at the highest speed to ensure cell lysis and/or release of DNA from soil particles.
5. If using Bulat buffer, incubate overnight with rotation in an oven set to 65°C. If using PowerMax® Bead Solution, proceed to step 6.
6. Spin the 50-mL tube at 2,500×*g* for 3 min and transfer the supernatant to a clean tube containing 5 mL of C3 (see Note 4).
7. Incubate at 4°C for 10 min to aid precipitation of non-DNA organic and inorganic materials, humic substances, cell debris, and proteins (see Note 5).
8. Spin the tube in the centrifuge at 2,500×*g* for 4 min, then transfer the supernatant to a clean tube containing 4 mL C3 solution.
9. Incubate at 4°C for 10 min.
10. Spin at 2,500×*g* for 4 min, then remove the supernatant to a clean 50-mL tube and add 30 mL of solution C4 (guanidine HCl—isopropanol solution).
11. Spin the resulting solution through silicon spin filters (see Note 6).
12. Add 10 mL of solution C5, an ethanol-based wash solution, to clean the DNA that is bound to the silica filter membrane in the spin filter (see Note 7).
13. Add 1.5–5 mL of solution C6 (Tris buffer solution) to the spin filter membrane and centrifuge at 2500×*g* in order to elute the bound DNA (see Note 8).
14. Transfer the eluate to 1.5-mL tube(s).
15. Store the DNA extract at −20°C.

3.2. Small Extraction

1. Place up to 0.5 g (wet weight) of samples into a 2-mL FASTPrep® tube containing 250-mg glass beads (see Notes 9 and 10).
2. Add 600 μL of Bulat extraction buffer.
3. Place the tube in a FastPrep® Instrument (QBIOgene) and shake for 45 s at speed 5.5. This causes samples to be pulverized and cells to be lysed.
4. Place samples on ice for 2 min.
5. Repeat steps 3 and 4 three times.
6. Place samples on a rotary mixer and incubate overnight with rotation at 65°C.
7. Adjust the mixture to 1.15 M NaCl and add 300 μL of chloroform/octanol (24:1).
8. Incubate at room temperature with rotation for 10 min.

9. Centrifuge at 12,000 × *g* for 2 min.
10. Remove the supernatant into a clean 1.5-mL tube.
11. Add 5× volume of PB-buffer (QIAGEN) to 1 volume of the supernatant and spin at 10,000 × *g* for 30–60 s and discard elute.
12. Add 0.5 mL of Salton wash 1 buffer (BIO 101).
13. Spin at 10 000 × *g* for 30–60 s and discard elute.
14. Add 0.5 mL of Salton wash 2 buffer (BIO 101).
15. Spin at 10 000 × *g* for 30–60 s and discard elute.
16. Add 0.5 mL of AW1 (QIAGEN).
17. Spin at 10 000 × *g* for 30–60 s and discard elute.
18. Add 0.5 mL of AW1 (QIAGEN).
19. Elute the DNA from the spin column into a clean 1.5-mL tube by spinning twice with 200 μL EB buffer (10 mM Tris–HCl, pH 8.5) (QIAGEN) 10 000 × *g* for 30–60 s.
20. Store the DNA extract at −20°C.

4. Notes

1. Proteinase K is an endolytic serine protease that cleaves after aliphatic, aromatic, and hydrophobic amino acids to break down protein structure (10). DTT reduces cystine cross-links in proteins to destroy their quaternary structure and allow further degradation. Sodium dodecyl sulfate (SDS) is a detergent which acts to denature proteins (e.g., nucleases) through interfering with noncovalent subunit interactions as well as solubilizing biological membranes (10). PTB cleaves glucose-derived protein cross-links (11) and has been shown to increase the success of PCR reactions from ancient material (12), although the exact mechanism by which it achieves this is unknown.
2. Bulat buffer often results in the final extract carrying less coextracts, but can lead to clogging of proteinaceous substances on the silica filters. It is best used with less organic-rich samples.
3. If solution C1 contains precipitates, heat at 60°C until the precipitate has dissolved.
4. Solution C3 is a second reagent to precipitate additional non-DNA organic and inorganic material including humic acid, cell debris, and proteins.
5. It is important to remove contaminating organic and inorganic matter that may reduce DNA purity and inhibit downstream DNA applications.

6. Solution C4 is a high concentration salt solution. Since DNA binds tightly to silica at high salt concentrations, this will adjust the DNA solution salt concentration to enable binding of DNA to the spin filters, but not non-DNA organic and inorganic material that may still be present at low levels.
7. This wash solution removes residual salt, humic acid, and other contaminants while allowing the DNA to stay bound to the silica membrane.
8. The DNA extract should be colorless. A dilution series using qPCR should be performed immediately after the extraction of DNA to assess any inhibition within the extract. If dilution does not resolve PCR inhibition, pass the extract through a 30,000 MWCO Millipore Amicon® ultracentrifuge tube, wash twice with ultrapure water, and elute in 200 μL of solution C6.
9. Each Lysing Matrix tube contains 1.4 mm ceramic spheres, 0.1 mm silica spheres, and one 4-mm glass bead.
10. It is important not to overload the FASTPrep® tubes. It is possible to extract larger volumes and combine the extracts at stage 10.

Acknowledgment

This work was supported by Murdoch University, Perth, Australia.

References

1. Haile J, MacPhee R, Roberts R, Arnold L, Brook B, Nielsen R, Gilbert M, Brock F, Munch K, Chivas A, Tikhonov A, Willerslev E (2009) Ancient DNA reveals late survival of mammoth and horse in interior Alaska. Proc Natl Acad Sci U S A 106:22363–22368
2. Haile J, Larson G, Owens K, Dobney K, Shapiro B (2010) Ancient DNA typing of archaeological pig remains corroborates historical records. J Archaeol Sci 37:174–177
3. Haile J, Holdaway R, Oliver K, Bunce M, Gilbert MTP, Nielsen R, Munch K, Ho S, Shapiro B, Willerslev E (2007) Ancient DNA chronology within sediment deposits: are paleobiological reconstructions possible and is DNA leaching a factor? Mol Biol Evol 24:982–989
4. Willerslev E, Hansen A, Binladen J, Brand T, Gilbert M, Shapiro B, Bunce M, Wiuf C, Gilichinsky D, Cooper A (2003) Diverse plant and animal genetic records from Holocene and Pleistocene sediments. Science 300:791–795
5. Lydolph M, Jacobsen J, Arctander P, Gilbert M, Gilichinsky D, Hansen A, Willerslev E, Lange L (2005) Beringian paleoecology inferred from permafrost-preserved fungal DNA. Appl Environ Microbiol 71:1012–1017
6. Crecchio C, Stotzky G (1998) Binding of DNA on humic acids: effect on transformation of *Bacillus subtilis* and resistance to DNase. Soil Biol Biochem 30:1061–1067
7. Khanna M, Stotzky G (1992) Transformation of *Bacillus subtilis* by DNA bound on montmorillonite and effect of DNase on the transforming ability of bound DNA. Appl Environ Microbiol 58:1930–1939
8. Gilbert MTP, Bandelt HJ, Hofreiter M, Barnes I (2005) Assessing ancient DNA studies. Trends Ecol Evol 20:541–544
9. Bulat S, Lubeck M, Alekhina I, Jensen F, Knudsen I, Lubeck P (2000) Identification of a universally primed-PCR-derived sequence-characterized amplified region marker for an antagonistic strain of *Clonostachys rosea* and

development of a strain-specific PCR detection assay. Appl Environ Microbiol 66:4758–4763
10. Voet D, Voet J (1995) Biochemistry. Wiley, New York
11. Vasan S, Zhang X, Zhang XN, Kapurniotu A, Bernhagen J, Teichberg S, Basgen J, Wagle D, Shih D, Terlecky I, Bucala R, Cerami A, Egan J, Ulrich P (1996) An agent cleaving glucose-derived protein crosslinks in vitro and in vivo. Nature 382:275–278
12. Poinar HN, Hofreiter M, Spaulding WG, Martin PS, Stankiewicz BA, Bland H, Evershed RP, Possnert G, Paabo S (1998) Molecular coproscopy: dung and diet of the extinct ground sloth *Nothrotheriops shastensis*. Science 281:402–406

Chapter 9

DNA Extraction from Fossil Eggshell

Charlotte L. Oskam and Michael Bunce

Abstract

Avian eggshell fragments recovered from both paleontological and archaeological deposits contain a cache of well-preserved ancient DNA. Here, we describe an extraction protocol that has been optimized to maximize the recovery of ancient DNA from fossil eggshell and minimize the co-purification of PCR inhibitors. In this method, fossil eggshell fragments are powdered, then digested and heated to release DNA from the calcite matrix. The digest then undergoes a concentration step before purification and washing using silica columns. The method has been used to recover aDNA from the eggshell of many avian species including moa, elephant birds, and emu, up to 19,000 years old.

Key words: Eggshell, Silica, Ancient DNA, DNA extraction

1. Introduction

Amino acids and stable isotopes recovered from fossil eggshells have been used extensively to reconstruct palaeodiets and geochronology (1–4). Recently, we demonstrated that fossil eggshells are also a source of well-preserved ancient DNA (5). As determined by confocal microscopy, DNA contained within the eggshell is protected in calcite due to its intracrystalline deposition within the eggshell matrix. This protection also provides a barrier to contaminating exogenous DNA: quantitative PCR (qPCR) results showed that moa eggshell had on average 125 times less microbial DNA than moa bone, making it an attractive substrate for high-throughput sequencing applications (5). In addition, the aDNA within eggshell has been shown to persist in a wide range of climatic conditions and has been amplified from eggshell fragments many thousands of years old (5).

In this chapter, we describe a DNA extraction protocol to isolate DNA from fossil eggshell fragments. Using qPCR to monitor DNA yields (See Chapter 16), we have optimized this protocol to

Beth Shapiro and Michael Hofreiter (eds.), *Ancient DNA: Methods and Protocols*, Methods in Molecular Biology, vol. 840, DOI 10.1007/978-1-61779-516-9_9,

maximize DNA recovery and to minimize the co-purification of PCR inhibitors. In this method, powdered eggshell fragments are incubated in a digestion buffer for up to 24 h, including a final heat step at 95°C, which we suspect aids in solubilization of the calcite and releases the DNA from the crystalline matrix. The DNA is then concentrated on 30,000 Da MWCO columns and purified using commercial silica spin columns.

2. Materials

All reagents should be stored according to manufacturers' requirements. Preparations should be carried out at room temperature unless indicated otherwise, using appropriate anti-contamination controls (e.g. filter-tipped pipettes, DNA-free consumables, etc.).

2.1. Eggshell Sampling

1. 10% bleach and 100% Ethanol (analytical grade).
2. Eggshell powdering equipment: either a Dremel tool (hand-held drill) and drill bits (part #114 or #191) (Racine, WI, USA) or fine grit sand paper and a mortar and pestle (see Notes 1 and 2).
3. Aluminium foil ($\sim 20 \times \sim 30$ cm^2).
4. 2.0-mL safelock tubes.
5. Electronic weighing scale.

2.2. Eggshell Digestion

1. Digestion buffer (700 μL per sample) containing final volumes of: 0.47 M EDTA (pH 8.0), 20 mM Tris (pH 8.0), 1% Triton X-100, 10 mM Dithiothreitol (DTT), 1 mg mL^{-1} proteinase K (see Notes 3 and 4).
2. 50-mL falcon tube (one per sample).
3. Oven with a rotary mixer, wheel, or similar device to keep samples constantly in motion during incubation steps, or thermal mixer (allows temperatures up to 95°C).
4. Parafilm.
5. Pipettes—P1000, P200, P20, and aerosol-resistant pipette tips.

2.3. Eggshell Extraction

1. 1.5-mL safelock tubes (one per sample).
2. Vivaspin columns (30,000 MWCO).
3. Qiagen Kit containing: Qiagen columns, PBi buffer (see Note 5), and EB buffer.
4. AW1 wash buffer.
5. AW2 wash buffer.
6. Table top centrifuge for 1.5- and 2.0-mL tubes capable of approximately 16,000 ×*g*).

3. Methods

Procedures should be carried out at room temperature unless otherwise specified. All surfaces and equipment should be cleaned with bleach and then ethanol to eliminate contamination. Always include extraction negative controls.

3.1. Eggshell Sampling

1. Prior to sampling, using either dremel tool (#114 or #191 drill bit) or sandpaper, lightly grind off the outer surfaces of the eggshell sample to remove debris (see Note 1).
2. Working on top of a clean piece of clean aluminium foil and using a clean drill bit, powder 50–100 mg of eggshell (see Note 2).
3. Transfer the eggshell powder from the foil to a pre-weighed 2.0-mL safelock tube for digestion. Weigh the tube to determine the exact amount of powder used in the DNA isolation procedure.

3.2. Eggshell Digestion

1. Add 7 μL DTT solution and 14 μL proteinase K solution per 679 μL stable digestion buffer (EDTA, Tris, Triton X-100) to make the active digestion buffer (see Note 3). Mix well. Seal with parafilm and place in a rotating oven (or thermal mixer) for 5 min at 55°C to dissolve Triton X-100 (see Note 4).
2. Add the active digestion buffer (700 μL) to a 2.0 mL Eppendorf containing 50–100 mg of eggshell powder and seal with parafilm (a good seal is essential). Gently vortex the tube to homogenize the digestion buffer with the eggshell powder. Incubate with rotation for 2–24 h at 55°C.
3. Increase the temperature of the oven or block to 95°C. Meanwhile, vortex the sample tubes for 20 s. Once the desired temperature has been reached, incubate samples for 10 min at 95°C. Vortex each sample and repeat the heating step again. Or repeat step 3. (see Note 6).
4. Allow tubes to cool to room temperature on a bench and remove parafilm.
5. Proceed to DNA purification (below).

3.3. DNA Purification: Silica Method

1. Following digestion, centrifuge the sample at 16,000×*g* for 2 min and ensure any remaining undigested eggshell has settled.
2. Collect the supernatant and transfer to a 30,000 MWCO Vivaspin 500 column (Sartorius Stedim Biotech, Germany; see Note 7).
3. Centrifuge the Vivaspin column with the supernatant at 16,000×*g* for 10–20 min, to concentrate supernatant to ~50 μL.

4. Transfer the concentrated supernatant to a new 2-mL tube and combine with at least 5 volumes of Qiagen Buffer PBi and vortex to mix (see Note 5).
5. Using a bench centrifuge, spin sample for 10 s and add to a Qiaquick column with attached collection tube.
6. Centrifuge Qiaquick column for 1 min at 16,000×*g*, and discard the flow-through.
7. Wash with 700 μL Qiagen wash buffer AW1 by centrifuging sample for 1 min, and discard the flow-through.
8. Wash with 700 μL Qiagen wash buffer AW2 by centrifuging sample for 1 min, and discard the flow-through.
9. To ensure all buffer components have been removed, centrifuge for an additional 1 min. Then place the Qiaquick column in a clean 1.5-mL tube with the lid removed.
10. To elute the DNA, add 60 μL (or a volume appropriate to the concentration of DNA required) of Qiagen elution buffer EB directly to the centre of the silica membrane. Wait 5 min prior to centrifugation to allow the DNA to elute off the silica.
11. Centrifuge for 1 min at maximum speed to collect the EB, now containing DNA.
12. Transfer to new 1.5-mL tubes (that have lids). The DNA is ready for downstream molecular biology analyses (see Note 8).

4. Notes

1. Thin eggshell is very fragile and susceptible to crumbling. We recommend using a dremel with drill bit for thicker (>0.7 mm) and sandpaper for thinner eggshell fragments (<0.7 mm) to remove the outer surfaces of the eggshell.
2. Following the removal of the outer surfaces, thin eggshell becomes more fragile. To ensure a homogenous powder, a mortar and pestle may be preferred as using a dremel may result in 'chipped off' pieces of thin eggshell.
3. Make up fresh for each digestion and discard unused solution. DTT and proteinase K are not stable when added to make an active digestion solution; it is for that reason that the active buffer is made fresh for each digestion.
4. Detergents (i.e. Triton X-100) at 4°C will precipitate out of solution in the digestion buffer. If this occurs, the digest buffer should be heated to allow the detergent to fully dissolve. SDS (1%) can be substituted for 1% Triton X-100; however, an

increased level of inhibition has been observed during qPCR experiments (see (5)).

5. We have noticed that eggshells that have been thermally modified (i.e. burnt archaeological eggshell) may alter the pH during the DNA-binding step to the silica. This is observed as a change in pH colour, from a preferred yellow (pH <7.5) to a light–dark purple (pH >7.5), therefore it is imperative that the pH indicator (now supplied separately) is added to the Qiagen buffer PB before use. The pH change can be overcome by the addition of ~2 μL of 3 M sodium acetate (pH 5.2) to the 5 volumes of Qiagen buffer PBi. As it may take a few seconds to see a colour change, be sure to mix thoroughly before adding additional sodium acetate.
6. Although not all of the powdered eggshell (calcium carbonate) will digest, this 95°C heat step is particularly important, as we believe the heat releases the DNA from the intracrystalline eggshell matrix. However, it should be noted that disruption of the DNA duplex at 95°C may cause problems with downstream applications that require adapter ligation such as HTS library builds.
7. The 30,000 MWCO Vivaspin 500 columns in this protocol serve two purposes. First, small molecules that act as PCR inhibitors are allowed to pass through the column while the DNA is retained. Second, the MWCO membrane acts to concentrate the DNA into a volume more appropriate for the silica-binding step. The vertical polyethersulfone (PES) membranes of the Vivaspin 500 column (Sartorius Stedim Biotech, Germany) are preferable to those that use horizontal membranes (often cellulose-based), as they are more resistant to blockage. Whatever MWCO membrane is used, large differences in the flow rate of different samples through the columns are still commonly observed.
8. Investigators should be cognisant that it is not uncommon for the DNA extract to contain inhibitors detrimental to PCR; therefore, we recommend that a dilution series using qPCR is performed directly after the extraction of DNA to assess any inhibition and the best level of dilution for further use of the extract.

Acknowledgments

MB was supported by the Australian Research Council as a Future Fellow (FT0991741). We thank Emma McLay, Morten Allentoft, Jayne Houston, and James Haile for helpful advice.

References

1. Higham T (1994) Radiocarbon dating New Zealand prehistory with moa eggshell: some preliminary results. Quat Sci Rev 13:163–169
2. Johnson BJ, Miller GH, Fogel ML, Beaumont PB (1997) The determination of the late Quaternary paleoenvironments at Equus Cave, South Africa, using stable isotopes and amino acid racemization in ostrich eggshell. Palaeogeogr Palaeoclimatol Palaeoecol 136:121–137
3. Miller GH, Beaumont PB, Deacon HJ, Brooks AS, Hare PE, Jull AJT (1999) Earliest modern humans in southern Africa dated by isoleucine epimerization in ostrich eggshell. Quat Sci Rev 18:1537–1548
4. Miller GH, Fogel ML, Magee JW, Gagan MK, Clarke SJ, Johnson BJ (2005) Ecosystem collapse in Pleistocene Australia and a human role in megafauna extinction. Science 309:287–290
5. Oskam CL, Haile J, McLay E, Rigby P, Allentoft ME, Olsen ME, Bengtsson C, Miller GH, Schwenninger JL, Jacomb C, Walter R, Baynes A, Dortch J, Parker-Pearson M, Gilbert MT, Holdaway RN, Willerslev E, Bunce M (2010) Fossil avian eggshell preserves ancient DNA. Proc Biol Sci 277:1991–2000

Chapter 10

Ancient DNA Extraction from Plants

Logan Kistler

Abstract

A variety of protocols for DNA extraction from archaeological and paleobotanical plant specimens have been proposed. This is not surprising given the range of taxa and tissue types that may be preserved and the variety of conditions in which that preservation may take place. Commercially available DNA extraction kits can be used to recover ancient plant DNA, but modifications to standard approaches are often necessary to improve yield. In this chapter, I describe two protocols for extracting DNA from small amounts of ancient plant tissue. The CTAB protocol, which I recommend for use with single seeds, utilizes an incubation period in extraction buffer and subsequent chloroform extraction followed by DNA purification and suspension. The PTB protocol, which I recommend for use with gourd rind and similar tissues, utilizes an overnight incubation of pulverized tissue in extraction buffer, removal of the tissue by centrifugation, and DNA extraction from the buffer using commercial plant DNA extraction kits.

Key words: Ancient DNA, Plant DNA, DNA extraction, CTAB extraction, PTB extraction

1. Introduction

DNA recovered from archaeological and paleobotanical plant remains can be used to infer plant domestication histories, the movement of crop plants, paleoecological community models, and plant demographic histories (1–6). Plant tissues preserved by desiccation or freezing are ideal for ancient DNA (aDNA) research, but even samples that have been subjected to anaerobic waterlogging and carbonization have been shown to contain amplifiable DNA (7–9).

There is currently no standard protocol for DNA extraction from ancient plant remains, due largely to the diversity of plant taxa and tissue types recovered in ancient deposits. Leaf tissue is a preferred material for modern DNA extraction, but is rarely available archaeologically. Seeds are often used in aDNA analyses and

Beth Shapiro and Michael Hofreiter (eds.), *Ancient DNA: Methods and Protocols*, Methods in Molecular Biology, vol. 840, DOI 10.1007/978-1-61779-516-9_10, © Springer Science+Business Media, LLC 2012

have been shown to yield DNA after desiccation, waterlogging, and carbonization (7–14). Seeds preserved by anaerobic soil conditions, while uncommon (e.g. 15), might also yield recoverable DNA. Other ancient plants tissues yielding DNA have included gourd rind (4), maize cobs (3), wood (16–18), and vegetative tissues such as peatmoss shoots and seagrass rhizomes (19, 20). Like seeds, pollen grains are adapted for viable DNA storage and protection during dispersal, and pollen grains recovered from lake sediment cores have yielded ancient DNA (5, 21). Plant DNA has also been successfully recovered from the inner surfaces of ceramics from an archaeological shipwreck, revealing the contents of ancient Greek amphorae (22). DNA isolation has been attempted from siliceous phytoliths extracted from archaeological soils, but has not been successful (23). Further experimental studies are necessary to understand whether this failure is related to the silicification processes in plants, the archaeological contexts from which phytoliths were taken for aDNA extraction, or the protocols used.

Isolation of DNA from plant tissues is complicated by the presence of abundant polyphenols, sugars, secondary compounds, and other potential PCR inhibitors. Standard DNA extraction protocols and commercial kits designed to overcome these obstacles are often suitable for use with ancient plant remains (12, 14, 17–19, 24, 25). Modifications to manufacturers' protocols are often required, however, to accommodate different tissue types and preservation conditions. Modifications range from changes in incubation time and temperature to the use of additional reagents to combat PCR inhibition.

DNA extraction protocols based on the strong detergent cetyltrimethyl ammonium bromide (CTAB) have been used with plants since the mid-1980s (26, 27). They can be used with small amounts of tissue, including very small single seeds, and are adaptable to samples of various composition and preservation. CTAB protocols used previously with ancient plant remains, including CTAB/DTAB variations, are described elsewhere (7, 11, 20, 24). A protocol developed for modern leaf tissue (28) and modified for use with single chenopod (*Chenopodium* sp.) seeds (14) is described here. This protocol successfully recovered up to 6 μg of bulk DNA from single modern seeds, from which PCR products of approximately 1,300 base pairs (bp) amplified easily. The yield was lower and fragment lengths shorter when working with ancient plants, but PCR products amplified successfully from several single-seed extractions. This protocol begins with tissue disruption and incubation in extraction buffer, followed by DNA extraction using chloroform and purification using isopropanol and ammonium acetate. DNA is then washed in ethanol and suspended in TE Buffer.

Extraction techniques that incorporate *N*-phenacylthiazolium bromide (PTB) may be appropriate when CTAB and other standard protocols are unsuccessful. PTB is believed to increase yield by releasing DNA from DNA–protein cross-links (29, 30). PTB has been used in coprolite DNA extractions to counteract the inhibitory effects of Maillard products, and also in extractions of various modern and ancient plant tissues (3, 4, 29, 31, 32). PTB-based extraction performs very well in modern wood extraction when compared with CTAB and unmodified Qiagen kit extractions (32). PTB extraction has been used to isolate DNA from desiccated maize cobs without attached seeds (3) and from modern and ancient bottle gourd rind tissue (4).

PTB extraction protocols for use with plants have been described previously (3, 32) and typically include a long incubation period in EDTA, followed by the addition of PTB and often proteinase K, phenol-chloroform extraction, and DNA recovery through precipitation or silica-binding. Erickson and others (4) developed a simple PTB-based protocol for use with bottle gourd rind. I provide a modified version of this protocol below, in which tissue disruption is followed by an overnight incubation in extraction buffer. The tissue is then centrifuged out of the mixture and DNA is extracted from the supernatant using a Qiagen DNEasy Plant Mini Kit (Qiagen).

2. Materials

2.1. CTAB Protocol

1. CTAB buffer: 2% (w/v) CTAB, 100 mM Tris–HCl pH 8.0, 20 mM EDTA, 1.4 M NaCl. 500 μL per sample.
2. β-Mercaptoethanol, 5 μL/1 mL of CTAB buffer.
3. Polyvinylpyrrolidone (PVP), 40 mg/1 mL of CTAB buffer.
4. Chloroform, 500 μL per sample.
5. Isopropanol, approximately 200–300 μL per sample, on ice.
6. 7.5 M Ammonium acetate, approximately 25–35 μL per sample, on ice.
7. 70% Ethanol, 700 μL per sample.
8. 95% Ethanol, 700 μL per sample.
9. TE buffer: 10 mM Tris–HCl pH 8.0, 1 mM EDTA. 50 μL per sample, more to dilute DNA if necessary.
10. Sterile pellet pestles.
11. 1.5-mL microcentrifuge tubes.
12. Water bath or heat block.
13. Table top centrifuge for 1.5/2-mL tubes capable of 13,000 × *g* RCF.
14. Fume hood rated for use with chloroform and β-mercaptoethanol.

2.2. PTB Protocol

1. PTB extraction buffer: 1% SDS, 10 mM Tris, pH 8.0, 5 mM NaCl. 50 mM DTT, 0.4 mg/mL proteinase K, 10 mM EDTA, 2.5 mM *N*-phenacylthiazolium bromide (PTB). 1.2 mL per sample.
2. Qiagen Plant DNEasy Mini Kits, 1 per sample. The lysis buffer (AP1) and RNaseA included with the kit are not necessary.
3. Shaker bath, or another means of incubating samples at 37°C with constant agitation.
4. Mechanized mill, such as a bead mill, or a sander wheel attachment on a power drill.
5. 1.5- and 2-mL microcentrifuge tubes.
6. Table top centrifuge for 1.5/2-mL tubes capable of 20,000 × *g* (see Note 1).

3. Methods

3.1. CTAB Protocol

See Notes 2–4 before beginning.

1. Preheat the water bath or heat block to 55°C.
2. Mix the extraction buffer by adding to the CTAB buffer: PVP to 1 mM (40 mg/mL) and β-mercaptoethanol to 0.5% (v/v, 5μL/mL) (see Notes 5–6). Mix gently by inverting and heat slightly to dissolve PVP if necessary. Do not shake the buffer, as the detergent will foam easily.
3. Soak one seed in 500 μL extraction buffer in a 1.5-mL tube for 1 h at 55°C, agitating periodically (see Notes 7–9).
4. Following the incubation period, grid the seed without removing it from the tube using a sterile pellet pestle, and vortex briefly to homogenize the mixture. No large clumps of tissue should remain (see Note 10).
5. Add 500 μL of chloroform to each tube and mix gently (see Note 11).
6. Centrifuge for 7 min (see Note 12).
7. Transfer the aqueous phase (top layer) from each tube into a new tube, taking care to leave behind the bottom layer and the debris-filled interface.
8. Estimate the volume of the transferred aqueous phase. Add 0.08 volumes of cold ammonium acetate and 0.54 volumes of cold isopropanol. Invert 20–30 times to mix (see Note 13).
9. Incubate the tubes on ice for at least 30 min and up to 1 h.
10. Centrifuge for 3 min.
11. Carefully discard the supernatant without disturbing the DNA pellet (if visible) (see Note 14).

12. Add 700 μL of 70% ethanol to each tube, invert 5–10 times, and centrifuge for 1 min.
13. Carefully discard the supernatant without disturbing the DNA pellet (if visible).
14. Add 700 μL of 95% ethanol, invert 5–10 times, and centrifuge for 1 min.
15. Carefully discard the supernatant without disturbing the DNA pellet (if visible).
16. Invert the tubes on a paper towel briefly (2–3 min) to eliminate most of the moisture. Then leave the tubes right side up, but covered with a clean paper towel or tissue, until thoroughly dried, at least 1 h or up to overnight.
17. Rehydrate the samples with 50 μL TE Buffer at room temperature overnight (see Note 15).

3.2. PTB Protocol

See Note 16 before beginning.

1. Preheat the shaker bath (or alternative) to 37°C.
2. Prepare the rind tissue by removing exposed tissue with a sterile razor blade or scalpel. Do not remove the tough outer rind (exocarp), as it tends to yield more DNA than the cork-like inner rind (mesocarp). Wipe the outer rind clean and lightly bleach it, taking care to thoroughly remove all bleach with ethanol and dry the surface completely before extraction.
3. Grind 0.1–0.2 g of rind tissue to a fine powder using a mechanized mill or sander wheel, sterilizing the grinding equipment thoroughly between uses (see Note 17).
4. Add the powder to 1.2 mL of PTB extraction buffer in a 2-mL or larger tube, and vortex to homogenize thoroughly. The mixture should be somewhat fluid, not a dry cake in the tube. Add more PTB extraction buffer, if necessary, to achieve the desired consistency.
5. Incubate the mixture at 37°C with constant agitation for 18–24 h.
6. Centrifuge the mixture at 9,000 × *g* for 5 min. The samples should separate into a dense mass of tissue and about 500–700 μL of supernatant. If the tissue is not suitably compacted (i.e. if more than a very small amount of visible debris is suspended in the supernatant), centrifuge for an additional 2 min at up to 16,000 × *g*.
7. Transfer the supernatant from each tube to a new 1.5- or 2-mL tube, and estimate the recovered volume.
8. Add 0.325 volumes of Qiagen Buffer AP2, mix, and incubate on ice 5 min.

9. Complete the extraction by following the manufacturer's protocol provided with the Qiagen kit, beginning with step 10 (see Notes 18–20).

4. Notes

1. 20,000 × *g* is recommended in the manufacturer's protocol for the Qiagen kits, but 13,000 × *g* is generally adequate.
2. This protocol was developed for chenopods, which are weedy dicots that produce small (1–2 mm diameter), starchy seeds. When working with monocots, note that monocot seeds store nutrients for the embryo in starchy endosperm that often comprises the bulk of the seed tissue (e.g. cereals), while dicots use large cotyledons for storage. Cotyledons form with normal ploidy and become crucial photosynthetic organs following germination, while monocot endosperm in seeds is strictly used for embryonic nourishment and forms at half the plant's normal ploidy (e.g. hexaploid breadwheat forms a large amount of triploid endosperm). Relative quantity and placement of plastids might also be important if cpDNA is being targeted.
3. Severe PCR inhibition is sometimes observed in specimens with a heavily lignified epidermal layer. Removal of this tissue prior to extraction increases PCR success dramatically. This type of protective maternal tissue can be removed with a sterile razor blade or scalpel to facilitate DNA extraction directly from the embryo, especially when PCR inhibition is observed.
4. This protocol is not effective with protein-rich, oily bottle gourd (*Lagenaria siceraria* (Molina) Standl.) seeds, even when paired with extra phenol-chloroform purification steps. Similar seeds such as sunflower and squash may also be unsuitable.
5. When these components have been added, the shelf life of the buffer is limited (2–3 days), so make only enough for immediate use.
6. All steps using β-mercaptoethanol or chloroform should be performed under a fume hood.
7. To use tissues other than seeds, grind 10–20 mg of desiccated material prior to incubation.
8. To increase yield using larger volumes of tissue, increase the amount of extraction buffer, taking care to use enough so that the plant tissue does not form a semi-solid cake in the tube. Increase all other reagents prior to TE rehydration proportionally. It may also be necessary to increase tube size.

9. An optional addition of 1.5 μL RNase A and 15-min incubation at 37°C can be included after step 4, but is typically not necessary when working with ancient samples.
10. For tough samples, add a small amount of sterile sand to assist with grinding. This may not be sufficient for very tough seeds. Grind very tough samples using a small mortar and pestle or a mechanized mill prior to incubation, taking care to thoroughly sterilize the equipment using a strong bleach solution (10%) between samples. For small seeds, try to avoid grinding outside the incubation tube to reduce loss of tissue. Degraded ancient samples might be fragile enough for grinding before step 3. The incubation in buffer helps slightly soften tough samples.
11. Pure chloroform or chloroform:isoamyl alcohol 24:1 may be used.
12. All centrifugation steps should be carried out at 13,000–16,000 × *g*.
13. The aqueous phase following chloroform extraction is typically 300–350 μL for small seeds.
14. It is very unlikely that small seeds, especially of ancient origin, will yield a visible DNA pellet.
15. DNA concentration may vary considerably and depends on taxon, tissue type, and sample preservation. Template volume in subsequent PCR and other applications should be optimized accordingly.
16. This protocol is highly effective with bottle gourd (*Lagenaria siceraria* (Molina) Standl.) rind tissue, yielding nuclear and chloroplast DNA that amplifies easily (4). It is recommended for similar tissues, but has not been shown to be effective with wood DNA extraction. DNA yields from wood are expected to be low, regardless of the extraction protocol used. Given that the secondary xylem undergoes programmed cell death, including the dissolution of organelles and nucleic acids, the only recoverable DNA in the woody matrix is likely found in the axial and radial parenchyma cells. To use PTB with wood samples, consider using the protocol described by Asif and Cannon (32). Alternatively, Qiagen kit modifications for wood DNA extractions are described elsewhere (16–18, 33). CTAB extraction is not recommended for wood samples.
17. To avoid overheating while grinding with a dremel or drill, grind slowly and monitor temperature closely. If heat becomes problematic, periodically soak the attachment in ice water and dry thoroughly before continuing.
18. A 1.5-mL tube will only accommodate 600 μL of lysate going into the kit's step 13. Use a larger tube if more lysate is recovered.

19. Several repetitions of the kit's step 14 may be necessary, depending on the volume of lysate recovered in step 12.
20. It is recommended to elute using a total of 100 μL of Qiagen Buffer AE to increase template concentration. Elution volume may be altered according to sample size and quality.

References

1. Schlumbaum A, Tensen M, Jaenicke-Despres V (2008) Ancient plant DNA in archaeobotany. Veg Hist Archaeobot 17(2):233–244
2. Gugerli F, Parducci L, Petit RJ (2005) Ancient plant DNA: review and prospects. New Phytol 166:409–418
3. Jaenicke-Deprés V et al (2003) Early allelic selection in maize as revealed by ancient DNA. Science 302:1206–1208
4. Erickson DL et al (2005) An Asian origin for a 10,000-year-old domesticated plant in the Americas. Proc Natl Acad Sci USA 102(51): 18315–18320
5. Parducci L et al (2005) Ancient DNA from pollen: a genetic record of population history in Scots pine. Mol Ecol 14:2873–2882
6. Gould BA et al (2010) Evidence of a high-Andean, mid-Holocene plant community: an ancient DNA analysis of glacially preserved remains. Am J Bot 97(9):1579–1584
7. Pollmann B, Jacomet S, Schlumbaum A (2005) Morphological and genetic studies of waterlogged Prunus species from the Roman vicus Tasgetium (Eschenz, Switzerland). J Archaeol Sci 32:1471–1480
8. Schlumbaum A, Neuhaus JM, Jacomet S (1998) Coexistence of tetraploid and hexaploid naked wheat in a Neolithic lake dwelling of Central Europe. Evidence from morphology and ancient DNA. J Archaeol Sci 25:1111–1118
9. Brown TA et al (1994) DNA in wheat seeds from European archaeological sites. Cell Mol Life Sci 50(6):571–575
10. Cappelini E et al (2010) A multidisciplinary study of archaeological grape seeds. Naturwissenschaften 97:205–217
11. Manen JF et al (2003) Microsatellites from archaeological *Vitis vinifera* seeds allow a tentative assignment of the geographical origin of ancient cultivars. J Archaeol Sci 30:721–729
12. Li C et al (2011) Ancient DNA analysis of desiccated wheat grains excavated from a Bronze Age cemetery in Xinjiang. J Archaeol Sci 38:115–119
13. Szabó Z et al (2005) Genetic variation of melon (*C. melo*) compared to an extinct landrace from the Middle Ages (Hungary). I. rDNA, SSR and SNP analysis of 47 cultivars. Euphytica 146:87–94
14. Kistler L, Shapiro B (2011) Ancient DNA confirms a local origin of domesticated chenopod in Eastern North America. J Archaeol Sci 38(12):3549–3554
15. Smith BD, Yarnell RA (2009) Initial formation of an indigenous crop complex in eastern North America at 3800 B.P. Proc Natl Acad Sci USA 106:6561–6566
16. Dumolin-Lapegue S (1999) Amplification of oak DNA from ancient and modern wood. Mol Ecol 8:2137–2140
17. Deguilloux MF et al (2006) Genetic analysis of archaeological wood remains: first results and prospects. J Archaeol Sci 33:1216–1227
18. Liepelt S et al (2006) Authenticated DNA from ancient wood remains. Ann Bot 98: 1107–1111
19. Suyama Y, Gunnarsson U, Parducci L (2008) Analysis of short DNA fragments from Holocene peatmoss samples. Holocene 18:1003–1006
20. Raniello R, Procaccini G (2002) Ancient DNA in the seagrass *Posidonia oceanica*. Mar Ecol Prog Ser 227:269–273
21. Bennett KD, Parducci L (2006) DNA from pollen: principles and potential. Holocene 16:1031–1034
22. Hansson MC, Foley BP (2008) Ancient DNA fragments inside Classical Greek amphoras reveal cargo of 2400-year-old shipwreck. J Archaeol Sci 35:1169–1176
23. Elbaum R et al (2009) New methods to isolate organic materials from silicified phytoliths reveal fragmented glycoproteins but no DNA. Quat Int 193:11–19
24. Elbaum R et al (2006) Ancient olive DNA in pits: preservation, amplification and sequence analysis. J Archaeol Sci 33:77–88
25. Russo EB et al (2008) Phytochemical and genetic analyses of ancient cannabis from Central Asia. J Exp Bot 59(15):4171–4182

26. Doyle JJ, Doyle JL (1987) A rapid DNA isolation procedure for small quantities of fresh leaf tissue. Phytochem Bull 19:11–15
27. Rogers SO, Bendich AJ (1985) Extraction of DNA from milligram amounts of fresh, herbarium and mummified plant tissues. Plant Mol Biol 5:69–76
28. Kalisz (2008) CTAB DNA Extraction Protocol. Univ. Pitt. http://www.pitt.edu/~kalisz/Protocols.html. Accessed 29 November 2010.
29. Poinar HN et al (1998) Molecular coproscopy: dung and diet of the extinct ground sloth *Nothrotheriops shastensis.* Science 281(402): 402–406
30. Rohland N, Hofreiter M (2007) Ancient DNA extraction from bones and teeth. Nat Protoc 2(7):1756–1762
31. Poinar HN (2002) The genetic secrets some fossils hold. Acc Chem Res 35(8):676–684
32. Asif MJ, Cannon CH (2005) DNA extraction from processed wood: a case study for the identification of an endangered timber species (*Gonystylus bancanus*). Plant Mol Biol Rep 23:185–192
33. Rachmayanti Y et al (2006) Extraction, amplification and characterization of wood DNA from Dipterocarpaceae. Plant Mol Biol Rep 24:45–55

Chapter 11

DNA Extraction from Formalin-Fixed Material

Paula F. Campos and Thomas M.P. Gilbert

Abstract

The principal challenges facing PCR-based analyses of DNA extracted from formalin-fixed materials are fragmentation of the DNA and cross-linked protein–DNA complexes. Here, we present an efficient protocol to extract DNA from formalin-fixed or paraffin-embedded tissues (FFPE). In this protocol, protein–DNA cross-links are reversed using heat and alkali treatment, yielding significantly longer fragments and larger amounts of PCR-amplifiable DNA than standard DNA extraction protocols.

Key words: Formalin, Formalin-fixed, Paraffin-embedded, DNA extraction, Ancient DNA

1. Introduction

The quality of DNA present in formalin-fixed material is generally poor, due both to fragmentation undergone by the DNA molecule itself, in particular when unbuffered formalin is used as a fixative, and to the formaldehyde-driven cross-linking of DNA with proteins (1–7). As a result, genetic analyses of such material must restrict themselves to targeting relatively short fragments of potentially amplifiable DNA. Ultimately, the quality of the DNA in a formalin-fixed sample is difficult to predict, as the degree of degradation and cross-linking results from a complex interplay of factors including the duration, pH, strength, and temperature of the fixative, plus the amount of time that has passed and the conditions of storage since fixation. As a general rule, long fixation times, in particular if longer than 12–24 h, highly acidic fixatives, longer periods of time in storage, and storage in warm environments all contribute to DNA decay. Over time, these factors will lead to the destruction of any remaining DNA in the sample.

A large number of methods have been proposed to recover DNA from formalin-fixed material (for a review see ref. (8)). Given that it is currently difficult, if not impossible, to reverse the

Beth Shapiro and Michael Hofreiter (eds.), *Ancient DNA: Methods and Protocols*, Methods in Molecular Biology, vol. 840,
DOI 10.1007/978-1-61779-516-9_11,

effects of DNA fragmentation, in our opinion, the key steps of successful methods are those that act through breakage of the DNA–protein cross-links using either heat or alkali treatment, or both. In this chapter, we report a modification of a method (9, 10) that we have found previously to be both simple and extremely effective in this regard. The method involves an initial incubation of the fixed tissue in a hot alkali buffer, coupled with a subsequent organic purification of the nucleic acids. The method results in similar total yields of DNA per sample (as measured in μg/μL) compared to other extraction methods such as commercial kits marketed with the specific aim of processing formalin-fixed or paraffin-embedded (FFPE) material. However, the method increases both the length of PCR-amplifiable fragments and the quantity of long-fragment DNA, as measured through conventional and quantitative real-time PCR (8–10).

2. Materials

Prepare the digestion buffer using molecular biology grade reagents at room temperature, using appropriate anti-contamination controls (e.g., filter-tipped pipettes, DNA-free consumables, etc.). To minimize the risk of contamination, we recommend purchasing ready-made solutions, as opposed to making them in the laboratory.

1. Alkali digestion buffer: 0.1 M NaOH with 1% SDS solution. Store at room temperature. The pH should be around 12.
2. 25:24:1 phenol:chloroform:isoamyl alcohol.
3. Chloroform.
4. Isopropanol.
5. 3 M sodium acetate, approximate pH 5.
6. (*Optional*) DNA precipitation "carrier," e.g., Glycoblue (Ambion, Inc., Austin, TX).
7. Molecular Biology Grade ethanol, 85%.
8. TE elution buffer: 10 mM Tris–HCl, 1 mM EDTA (pH 8.0).
9. Microtome with sterile (DNA-free) blades or sterile scalpel blades.
10. Sterile 2-mL O-ring screw-top tubes.
11. Sterile 1.5-mL centrifuge tubes.
12. Centrifuge(s) for 1.5/2-mL tubes (>10,000 × *g*).
13. Autoclave capable of heating to 120°C (preferable). Alternatively, water bath or hotblock heated to 100°C.
14. Tabletop vortex.

3. Methods

Carry out all procedures at room temperature unless otherwise specified. Always incorporate extraction blanks into the analysis. This protocol is suitable for either paraffin-embedded or non-embedded, formalin-fixed material (see Note 1).

3.1. Tissue Pre-Preparation

1. Obtain small sub-samples of tissue. For paraffin-embedded tissue, use a microtome to obtain several slices of tissue between 3 and 10 μm thick, or shave thin slices using a sterile scalpel blade (see Note 1). For non-paraffin-embedded tissue, obtain thin slices with a sterile scalpel blade.

3.2. Tissue Digestion

1. Place tissue in 0.5 mL of the alkali digestion buffer in a 2-mL screw-cap O-ring tube (see Notes 2 and 3).
2. Use an autoclave to heat the tissue-buffer to 120°C for 25 min. If autoclave use is not convenient, heating on a heat block or in a boiling water bath at 100°C for 40 min is an alternative (see Note 4).
3. Allow the tissue-buffer to cool for 5 min to room temperature (see Note 5).
4. Add 500 μL 25:24:1 phenol:chloroform:isoamyl alcohol to the mixture (see Note 6).
5. Agitate gently at room temperature for 5 min.
6. Centrifuge for 5 min at >10,000 × *g* to separate the layers.
7. Carefully remove the upper aqueous layer and add to a new tube containing 500 μL chloroform. Be careful not to remove the protein-containing interface. Discard the lower phenol layer (see Note 6).
8. Repeat steps 5–6 in Subheading 3.2.
9. Remove the upper aqueous layer and place in a new 1.5 mL eppendorf tube. Discard the lower chloroform layer (see Note 6).
10. Add 0.6–1 volume isopropanol and 0.1 volume 3 M sodium acetate (approx. pH 5). A small amount of commercial carrier solutions can also be added if required to facilitate pellet visualization, such as Glycoblue (Ambion, Inc., Austin, TX), following the manufacturers' guidelines. Mix well (see Note 7).
11. Immediately centrifuge at high speed (>10,000 × *g*) for 30 min at room temperature.
12. Immediately following centrifugation, decant the liquid from the tube carefully. The DNA will have precipitated into a pellet at the bottom of the tube and may not be visible.

13. To rinse the pellet, gently add 500–1,000 μL 85% ethanol, gently invert once, then centrifuge for 5 min at high speed.
14. Gently decant the ethanol. Repeat if necessary.
15. All ethanol must be removed from the pellet as any residual ethanol will inhibit downstream applications. This can be easily achieved with a small bore pipette, followed by a brief incubation at a relatively high temperature (e.g., 55–75°C).
16. Re-suspend the pellet in a suitable volume of TE buffer or ddH_2O (e.g., 50–100 μL). If the pellet has become very dry, leave it at room temperature in the liquid for 5–10 min, followed by gentle pipetting.

4. Notes

1. It is not necessary to use a solvent to remove paraffin from paraffin-embedded samples prior to extraction; however, if present in large amounts, it is helpful to trim it away with a sterile scalpel first.
2. The use of O-ring screw-cap tubes is extremely important, as under subsequent heating, high pressure will build up in the tube. Lids on tubes without an O-ring seal and screw fitting will be blown open.
3. Once added, the alkali digestion solution will begin to degrade the DNA, thus delays in the subsequent steps (up to precipitation) should be avoided.
4. It takes time to heat up and cool down an autoclave, thus 25 min should represent the time at 120°C and not the entire time in the autoclave. The use of cooler temperatures (100°C) is not as effective as the original protocol, but nevertheless yields significant improvements over other methods that do not incorporate a heat step.
5. The tissue will not have fully dissolved. This does not affect the results.
6. Organic extractions use phenol and chloroform to help purify the DNA. Both phenol and chloroform are toxic, and phenol in particular is extremely dangerous. Neither should be used without appropriate local training. Always handle both liquids and their containers with extreme care, using appropriate face, hand, and body protection. Do not handle using latex gloves as these are permeable to phenol and chloroform; use only nitryl gloves. The fumes of both are dangerous, thus always manipulate in a vented fume hood. Disposal of both requires conformation to specific regulations, thus relevant local disposal regulations should be consulted.

7. Isopropanol precipitation is most effective at relatively high centrifugal forces and in small tubes with a pointed end (the area covered by the precipitated DNA that forms the observed DNA pellet is most concentrated and thus easiest to spot and re-suspend if 1.5-mL tubes or smaller are used).

Acknowledgments

MTPG was supported by the Danish National Science Foundation's "Skou" grant program.

References

1. Brutlag D, Schlehuber C, Bonner J (1969) Properties of formaldehyde-treated nucleohistone. Biochemistry 8:3214–3218
2. Varshavsky A, Sundin O, Bohn M (1979) A stretch of "late" SV40 viral DNA about 400 bp long which includes the origin of replication is specifically exposed in SV40 minichromosomes. Cell 16:453
3. Ilyin Y, Georgiev G (1969) Heterogeneity of deoxynucleoprotein particles as evidencec by ultracentrifugation of cesium chloride density gradient. J Mol Biol 41:299
4. Feldman M (1973) Reactions of nucleic acids and nucleoproteins with formaldehyde. Prog Nucleic Acid Res Mol Biol 13:1–49
5. Varshavsky A, Ilyin Y (1974) Salt treatment of chromatin induces redistribution of histones. Biochim Biophys Acta 340:207–217
6. Jackson V (1978) Studies on histone organization in the nucleosome using formaldehyde as a reversible cross-linking agent. Cell 15:945–954
7. Møller K, Rinke J, Alexander R et al (1977) The use of formaldehyde in RNA-protein cross-linking studies with ribosomal subunits from *Escherichia coli*. Eur J Biochem 76:175–187
8. Gilbert M, Haselkorn T, Bunce M et al (2007) The isolation of nucleic acids from fixed, paraffin-embedded tissues-which methods are useful when? PLoS One 2:537
9. Shi SR, Cote RJ, Wu L et al (2002) DNA extraction from archival formalin-fixed, paraffin-embedded tissue sections based on the antigen retrieval principle: heating under the influence of pH. J Histochem Cytochem 50:1005–1011
10. Shi SR, Datar R, Liu C et al (2004) DNA extraction from archival formalin-fixed, paraffin-embedded tissues: heat-induced retrieval in alkaline solution. Histochem Cell Biol 122:211–218

Chapter 12

Case Study: Ancient DNA Recovered from Pleistocene-Age Remains of a Florida Armadillo*

Brandon Letts and Beth Shapiro

Abstract

Warm, humid regions are not ideal for long-term DNA preservation. Consequently, little ancient DNA research has been carried out involving taxa that lived in, for example, tropical and subtropical regions. Those studies that have isolated ancient DNA from warm environments have mostly been limited to the most recent several thousand years. Here, we discuss an ancient DNA experiment in which we attempt to amplify mitochondrial DNA from remains of armadillo, glyptodont, and pampathere from sites in Florida, USA, all believed to be around 10,000–12,000 years old. We were successful in recovering DNA from only one of these samples. However, based on the amount and distribution of DNA damage, the ancient DNA recovered was well-preserved despite the age and preservation environment. In this case study chapter, we discuss the experimental procedure we used to characterize the DNA from the Floridian samples, focusing on challenges of working with ancient specimens from warm environments and steps taken to confirm the authenticity of the recovered sequence.

Key words: Ancient DNA Extraction, Armadillo, Dasypus bellus, Mitochondrial DNA, Degraded DNA , Mefford Cave, Florida, Pleistocene

1. Introduction

Few ancient DNA (aDNA) studies have focused on Pleistocene-age animals that inhabited warm regions (1). This is due in part to the poor preservation of such samples compared to those preserved in colder, temperature-stable environments (2, 3). Remains from

* *Note*: In the case study presented in this chapter, we describe DNA extraction and amplification from ancient armadillo samples from Florida using a method similar to that presented in Chaps. 3 and 14. Other DNA extraction methods, such as the phenol:chloroform method described in Chap. 2, would also be appropriate for this type of sample. We discuss specific challenges associated with the analysis of ancient bone samples from warm regions.

Beth Shapiro and Michael Hofreiter (eds.), *Ancient DNA: Methods and Protocols*, Methods in Molecular Biology, vol. 840,
DOI 10.1007/978-1-61779-516-9_12,

areas such as Florida, where the climate is both warm and humid, are expected to decay quickly and present considerable challenges to the extraction and amplification of aDNA.

Pleistocene cingulates (armadillos, glyptodonts, and pampatheres) inhabited temperate to warm climates (4, 5). The remains of Pleistocene armadillos are dispersed mainly across the gulf coastal plain of North America, but have been found as far north as Missouri, Tennessee, and Nebraska (6). As they are common in these Pleistocene deposits, the cingulates are ideal to explore DNA survival in Pleistocene samples from warm, even subtropical areas. One reason for this abundance is that, in addition to skeletal components, each individual has around 1,000 osteoderms, which are the small bones that make up the carapace or shell. This results in a much larger number of preserved remains per individual, and therefore, a greater probability that some remains will be preserved. Additionally, the variety of deposition sites (open sites, river banks, submerged river bottoms, caves) where they are found makes it possible to compare DNA yield between preservation microenvironments.

2. Materials and Methods

We obtained 17 armadillo, glyptodont, and pampathere samples collected from various locations in Florida that are now part of the University of Florida Museum of Natural History collection. Samples were identified as belonging to *Dasypus bellus*, *Glyptotherium floridanum*, or *Holmesina septentrionalis*. All samples were estimated to be Rancholabrean in age, or about 10,000–12,000 years old. We performed ancient DNA extraction and PCR set-up at the Pennsylvania State University in a sterile, positive-pressure ancient DNA laboratory that is spatially isolated from modern molecular biology research. Workflow was always from the ancient DNA laboratory to the modern DNA laboratory, and full protective coverings were worn at all times. Negative controls were used at all steps, and PCR products were cloned to characterize DNA damage and identify environmental contaminants.

Before subsampling, we cleaned the outer surface of each bone around the subsampling site using a Dremel tool equipped with a cutting disk. This removes preservative coatings and limits potential contamination by exogenous sources such as human handling. As much as possible, care was taken to avoid the destruction of morphologically informative parts of the bones.

We removed subsamples from each bone using a Dremel tool equipped with either a drill tip or cutting disk. We collected powder from less dense samples by drilling directly into the interior of the bone. Drilling was the preferred method of subsampling due to its lower destructiveness; this process resulted in only a 2-mm hole and no other visible damage. We powdered bone fragments using a mikrodismembrator (Braun) by shaking at 600 rpm for 30 s–5 min, depending on the sample. For each specimen, we processed a final mass of 400–500 mg of bone powder.

We extracted DNA using the silica-based method described in Chap. 3. Darkly stained samples from river sites required modification of the protocol to repeat the wash step until the silica became free of discoloration (two or three repeated wash steps depending on the sample). We eluted the DNA in 50 μL of TE buffer.

We first attempted PCR amplification of the conserved mitochondrial 12S rDNA fragment from each extracted sample. We designed primers based on the sequence of the extant armadillo, *Dasypus novemcinctus,* as obtained from Genbank. We amplified a 97-base pair (bp) fragment of 12S using the primers Xen12S-56F, 5′-ATCAGCACACCAGTGAGAATG-3′; Xen12S-153R, 5′-GAGCAAAGCGTTGTGAGCTAC-3′.

In addition to amplifying 12S rDNA, we designed five overlapping primer sets to span 581 bp of the mitochondrial hypervariable region sequence that had been sequenced previously for modern *Dasypus* (7). We tested the primers using a modern individual, but, due to the high variability within this genomic region, had only limited success: only the 3′-most primer set resulted in amplification. We attempted to optimize the experiment by amplifying fragments of progressively increasing length, beginning with the 3′-most primer and pairing it with reverse primers from the other primer sets. This optimization (progressive amplification of longer fragments) was only ever performed using the ancient sample, so that no long fragments of amplified DNA were ever produced from the modern individuals and any resulting sequence is therefore unlikely to be that of a modern contaminant.

We performed PCR amplifications in 25 μL reactions consisting of 50 μg rabbit serum albumin, 0.25 mM dNTPs, 1× High Fidelity buffer, 1.25 units Platinum *Taq* High Fidelity (Invitrogen), 2 mM $MgSO_4$, 1 μM of each primer, and 1 μL DNA extract. Cycling conditions were 94°C for 60 s, followed by 50 cycles of 94°C for 30 s, 57°C (12S primers) or 50°C (control region primers) for 45 s, and 68°C for 45 s. No final extension was used. We cloned and sequenced four PCRs using the TOPO TA cloning kit (Invitrogen) in 1/10 reactions and BigDye terminator sequencing kit (Applied Biosystems) in 1/32 reactions. To create a consensus sequence, we aligned the resulting products using the Lasergene software suite (DNAstar, Inc.).

3. Results and Discussion

3.1. Sample Preservation

Only one of the 17 samples from which extraction was attempted yielded DNA. The sample was a tibia fragment from a Beautiful Armadillo, *Dasypus bellus* (UF 2478), from Mefford Cave, a limestone cavern in central Florida. The associated skeleton is the most complete that has been found and included within the carapace the skeletons of unborn offspring. The sample was extremely dry and brittle, and when powdered, was comparable to talcum powder. The exterior of the bone was brown and mottled in appearance, but the interior was a creamy off-white color.

We performed two extractions from this sample using powder taken from opposite ends of the bone. This made it possible to determine whether DNA was preserved throughout the sample, compare levels of damage across the specimen, and assess the authenticity of the sequence in an independent extraction. DNA extraction attempted from an associated osteoderm from the site yielded no amplifiable DNA. This could suggest that preservation varied between different parts of the skeleton. It is also possible that bone structure may affect the preservation of DNA: osteoderms, which function as armor, are small, dense, and easily fossilized (8). An osteoderm from a river site also failed to yield DNA; however, a test extraction of three 20-year-old nine-banded armadillo osteoderms revealed that DNA is present in modern osteoderms.

3.2. Troubleshooting the Experiment

We cloned three control region PCR products from the two extractions of the tibia and sequenced 39 clones. We identified 13 singleton substitutions: three C→T/G→A changes resulting from cytosine deamination; a transversion (A→C/T→G), most likely due to a polymerase misincorporation at an apurinic/apyrimidinic site (9); and nine A→G/T→C changes, which have also been shown to result from polymerase misincorporation in some ancient DNA samples (10). The two extractions yielded identical consensus sequences.

The 5′ end of the mitochondrial control region in the armadillo is highly repetitive (7). We found that this repetitive structure extended throughout the control region, making it difficult to design primers that would not bind in multiple places. Consequently, the products of each PCR comprised multiple, overlapping fragments that varied in length (Fig. 1). Because of the degraded nature of the specimens, it was not possible to circumvent this problem by designing longer fragments to span the repetitive sequence. We therefore chose to determine the control region sequence by cloning the PCR amplifications. This allowed us to separate the overlapping fragments and align them for a consensus sequence.

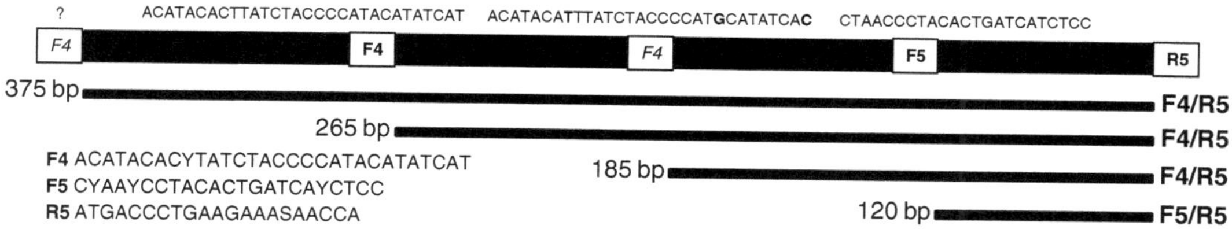

Fig. 1. Diagram showing the amplification products from the Mefford Cave armadillo specimen. A highly repetitive control region in *Dasypus* results in primers binding in multiple places. Primer F4 binds in at least three places, represented by *white boxes*. The correct binding site produces a 265-bp fragment and is highlighted with *bold text*. The unintended binding sites are indicated with *boxes* containing *italicized text*. The actual F4 binding site sequence for the 265- and 185-bp fragments is indicated above the corresponding box. Mismatches in the 185-bp fragment priming site are in *bold*. The binding site sequence for F5 is also provided. Primer sequences are provided below the diagram.

3.3. Challenges of a Warm Environment

Warm, wet environments are expected to dramatically increase the thermal age of ancient DNA, resulting in extensive damage and fragmentation (2). Our results suggest that Pleistocene-age DNA remains in some specimens were preserved in subtropical regions. However, it should be noted that the sample from which DNA was recovered had been preserved in a cave microenvironment. Cave interiors provide highly stable environments with little annual fluctuation in temperature or humidity and are known to promote the preservation of DNA. Mefford Cave may be exceptional among caves for long-term preservation of DNA: the current entrance is recent, and the small opening through which animal remains entered (presumably by washing in) during the Pleistocene has since been closed (11). Therefore, it is conceivable that Mefford Cave was sealed to the outside environment for an extended period of time following the deposition of this specimen.

References

1. Ramakrishnan U, Hadly EA (2009) Using phylochronology to reveal cryptic population histories: review and synthesis of 29 ancient DNA studies. Mol Ecol 18:1310–1330
2. Smith CI, Chamberlain AT, Riley MS, Stringer C, Collins MJ (2003) The thermal history of human fossils and the likelihood of successful DNA amplification. J Hum Evol 45: 203–217
3. Mitchell D, Willerslev E, Hansen A (2005) Damage and repair of ancient DNA. Mutat Res 571:265–276
4. Klippel W, Parmalee P (1984) Armadillos in North American late Pleistocene contexts. Spec Publ Carnegie Mus Nat Hist 8:149–160
5. Gillette DD, Ray CE (1981) Glyptodonts of North America. Smithsonian Contrib Paleobiol 40:1–262
6. Voorhies MR (1987) Fossil Armadillos in Nebraska: the Northernmost Record. Southwestern Nat 32:237–243
7. Huchon D, Delsuc F, Catzeflis FM, Douzery EJP (1999) Armadillos exhibit less genetic polymorphism in North America than in South America: nuclear and mitochondrial data confirm a founder effect in *Dasypus novemcinctus* (Xenarthra). Mol Ecol 8:1743–1748
8. Hill RV (2006) Comparative anatomy and histology of xenarthran osteoderms. J Morphol 267:1441–1460
9. Eckert KA, Kunkel TA (1991) DNA polymerase fidelity and the polymerase chain reaction. PCR Methods Appl 1:17–24
10. Stiller M, Green RE, Ronan M, Simons JF, Du L, He W, Egholm M, Rothberg JM, Keats SG, Ovodov ND, Antipina EE, Baryshnikov GF,

Kuzmin YV, Vasilevski AA, Wuenschell GE, Termini J, Hofreiter M, Jaenicke-Despres V, Paabo S (2006) Patterns of nucleotide misincorporations during enzymatic amplification and direct large-scale sequencing of ancient DNA. Proc Natl Acad Sci USA 103: 13578–13584

11. Auffenberg W (1957) A note on an unusually complete specimen of *Dasypus bellus* (Simpson) from Florida. Q J Fla Acad Sci 20:233–237

Chapter 13

Nondestructive DNA Extraction from Museum Specimens

Michael Hofreiter

Abstract

Natural history museums around the world hold millions of animal and plant specimens that are potentially amenable to genetic analyses. With more and more populations and species becoming extinct, the importance of these specimens for phylogenetic and phylogeographic analyses is rapidly increasing. However, as most DNA extraction methods damage the specimens, nondestructive extraction methods are useful to balance the demands of molecular biologists, morphologists, and museum curators. Here, I describe a method for nondestructive DNA extraction from bony specimens (i.e., bones and teeth). In this method, the specimens are soaked in extraction buffer, and DNA is then purified from the soaking solution using adsorption to silica. The method reliably yields mitochondrial and often also nuclear DNA. The method has been adapted to DNA extraction from other types of specimens such as arthropods.

Key words: Ancient DNA, Arthropods, Bones, Teeth, Museum specimens, Silica

1. Introduction

The research field of ancient DNA is generally accepted to have started in 1984, with the publication of short mitochondrial (mt) DNA fragments from the extinct quagga (1). However, it is important to note that the samples investigated in this study were a mere 140 years old, a typical age for many museum specimens of extant species. Since then, the number of studies using museum specimens for genetic investigations has risen sharply, be it for phylogenetic (e.g. (2–8)), phylogeographic, (e.g. (9–14)) or population genetic analyses (e.g. (15–20)). Sometimes, even studies on human genetic diversity (21) or paternity analyses of animal populations (22) rely on museum specimens.

This rising demand of molecular biologists to sample museum specimens is putting an increasing pressure on the collections of

Beth Shapiro and Michael Hofreiter (eds.), *Ancient DNA: Methods and Protocols*, Methods in Molecular Biology, vol. 840,
DOI 10.1007/978-1-61779-516-9_13,

natural history museums. Although a variety of tissue types can be used for genetic analyses (see (23) for a review), including hair (e.g. (24)), skin (e.g. (11)), or bird toe pads (e.g. (25)), the most common tissues used are bony specimens. However, while many studies yield exciting results, "consumptive sampling" (26), i.e., the removal and destruction of parts of the specimen, often irreversibly damages specimens and is in the long run unsustainable. As many specimens housed in natural history museums are from now-extinct populations or species (23), their preservation for future morphological as well as genetic studies is vital. Therefore, less destructive methods for DNA sampling have been developed, such as sampling from maxilloturbinal bone material (i.e. "the thin bones attached anteriorly to ridges inside the nasal cavity," (26)), a part of the skeleton that is not used for morphological studies. However, the ideal sampling method does not require any consumption of material, but rather preserves the morphological characters for future studies.

The method described below has been developed with exactly this aim, i.e., obtaining sufficient DNA for genetic analyses from bones and teeth without affecting the morphology of the specimens studied (27). Although it was initially developed for mtDNA analyses, subsequent studies (6, 8) have shown that many samples yield sufficient DNA to also allow analysis of nuclear DNA, at least up to a length of around 250 base-pairs (bp) (6). The method involves incubation of whole bone or teeth specimens in the extraction buffer for one to several days, followed by DNA recovery from the incubation solution using adsorption to silica in the presence of a chaotropic salt (generally guanidinium isothiocyanate, GuSCN). After extraction, samples are washed in double-distilled water to remove any traces of the extraction buffer and air-dried.

This treatment has no visible effect on the morphology of solid bone specimens (apart from them looking cleaner after the extraction; see Figs. 2 in (6, 27)), but very fragile specimens such as jaws or rostra from small mammals such as golden moles (8) may show signs of bone dissolution on the surface of the specimens.

The protocol described below is based on the initial publication of the method on bones and teeth (27). However, the method has been used for the extraction of DNA from arthropod specimens using both the original buffer conditions (28) and modified conditions (29, 30). In one of these studies (30), DNA was extracted from beetles up to 26,000 years old. Similarly, while the protocol below describes a silica-batch method for DNA purification, depending on the extraction buffer used, other DNA purification methods may be considered (31).

2. Materials

Prepare all solutions using HPLC grade water or water with a similar purity grade. Both the extraction and the binding buffer as well as the silica suspension are stable for at least 1 month. The washing buffer and the TE for elution are stable for several months.

1. Extraction/binding buffer: 5 M guanidinium isothiocyanate, 50 mM Tris–HCl, pH 8.0, 25 mM NaCl, 1.3% Triton-X100, 20 mM EDTA, 50 mM DTT (see also Notes 1–3).
2. Silica suspension: Weigh 4.8 g of silicon dioxide (recommended: Sigma-Aldrich, catalog number: S5631), add ddH_2O to 40 mL, and vortex until the silica is completely in suspension. Allow to settle for 1 h, transfer upper 39 mL into fresh tube, and allow to settle for another 4 h. Discard the upper 35 mL, leaving 4 ml of suspension/pellet, and add 48 μL 30% HCl. Vortex, aliquot, and store at room temperature in the dark.
3. Washing buffer 1: 5 M Guanidinium thiocyanate, 0.3 M sodium acetate (pH 5.2); store at RT in the dark.
4. Washing buffer 2: 50% Ethanol, 125 mM NaCl, 10 mM Tris–HCl, 1 mM EDTA (pH 8.0); store at RT.
5. Elution buffer (TE): 10 mM Tris–HCl, 1 mM EDTA (pH 8.0).
6. Rotary mixer, wheel, or similar device to keep samples constantly in motion during incubation steps.
7. Table top centrifuge for 1.5/2-mL tubes going up to 12,000 rpm.

3. Methods

All steps are performed at room temperature.

3.1. Incubation

1. Obtain an appropriate sample for extraction. For small species such as rodents, tenrecs, or insectivores, complete bones such as jaws or rostra can be used. For larger species, teeth are a good source, although when using incubation dishes of appropriate size, larger samples such as complete ape skulls can be extracted. In such cases, the extraction buffer volume needs to be adjusted accordingly, and DNA purification usually has to be done in multiple aliquots (see Notes 3 and 4).
2. When working in 15–50-mL tubes, add between 5 and 20 mL of extraction buffer to each sample. Seal tube with parafilm and incubate for 5 days under constant agitation in the dark (see also Notes 3–6).

3.2. DNA Purification

1. Either remove bone specimen from tube or transfer supernatant to a new tube.
2. Centrifuge the supernatant for 2 min at 12,000 × *g* to pellet any particles that have come off the sample. This is particularly important for samples that contain large amounts of dried soft tissue. Transfer as much of the liquid as possible into a new tube.
3. Add 100 μL of resuspended silica suspension and incubate for 3 h under constant movement in the dark (see Notes 6 and 7).
4. Centrifuge for 2 min at 5,000 × *g*, remove supernatant (see Note 8), and resuspend the silica pellet in 1 mL washing buffer 1 (see Note 9).
   At this step, you can also resuspend the silica pellet in 0.4 mL washing buffer 1 and proceed from step 4 of Subheading 3.2 of Chap. 3.
5. Centrifuge for 2 min at 5,000 × *g*, discard supernatant, and resuspend the silica pellet in 1 mL washing buffer 2.
6. Repeat step 4.
7. Centrifuge for 2 min at 16,000 × *g* and discard supernatant (see Note 10).
8. To completely remove any remaining supernatant, centrifuge again for 30 s at 16,000 × *g*, and remove any remaining supernatant (see Note 11).
9. Air-dry the silica by leaving the tubes with open lids at RT for about 15 min.
10. Add 50 μL elution buffer to the silica pellet, resuspend by carefully pipetting up and down and stirring with the pipette tip until you have a homogenous suspension (see Note 12).
11. Incubate for 10 min with closed lid.
12. Centrifuge for 2 min at 16,000 × *g*, transfer supernatant to a new, labeled tube, preferably a 0.5-mL tube; aliquot extract if required (see Note 13).
13. You may want to repeat steps 10 and 11, but the DNA yields of the second elution are generally much lower, so pooling of both elutions will result in a lower DNA concentration of the final extract.

3.3. Sample Curation

To avoid any salts of the extraction buffer infiltrating the samples, after removal from the extraction buffer, transfer them to a tube with double-distilled water. Incubate them overnight at RT, transfer them to a new tube with double-distilled water, and incubate for another few hours. Remove them from tube and let them air-dry slowly at room temperature.

4. Notes

1. When using fragile specimens such as bones from small mammals, it is advisable to adjust the extraction buffer, either by reducing the volume or the concentration of EDTA, which has a dissolving effect on bone. It is also possible to use a completely different buffer for extraction, as has been done as an adaptation of this method for DNA extraction from beetle specimens (29).
2. Although the initial study gave the best results using the GuSCN buffer, two other buffers (one Tris–NaCl-based, the other one sodium-phosphate-based) also yielded results with teeth (27).
3. Recent studies have shown that the optimal GuSCN concentration for maximizing DNA yields is around 1.4–1.7 M, rather than 5 M (31). However, it is unknown how a reduced GuSCN concentration affects the efficacy of DNA release. As other buffers have also yielded DNA using this method, it may be worthwhile to test the lower GuSCN concentration, especially when working with fragile specimens. Alternatively, after removal of the sample, you may want to dilute the extraction buffer to 1.5 M GuSCN using TE before adding silica. If nonchaotropic extraction buffers are used (e.g. (29)), add washing buffer 1 (this can also serve as binding buffer) in a ratio of 2 volumes extraction buffer to one volume washing buffer 1 after removal of the specimen and proceed from step 2 (see also Note 4).
4. The volume of extraction buffer needs to be adjusted depending on sample size. Tubes or dishes should be large enough to allow samples to move freely within them. Ideally, buffer should flow over the specimen during the agitation; avoid using too little buffer or filling up the tubes completely. If using more than 10 mL of extraction buffer, adjust the volume of silica suspension used for DNA binding. Volumes above 50 mL have to be purified in several parallel tubes. When using nonchaotropic salts, it is possible to concentrate the extraction buffer before silica purification using filter systems like the Vivaflow system (32). However, note that chaotropic salts destroy the filter membranes. If using nonchaotropic extractions buffers in combination with silica purification, adjust the volume of the GuSCN buffer (washing buffer 1) added for binding so that the ratio of extraction to binding buffer is 2:1 ((31) and Chap. 3; see also Note 6).
5. Rotation during incubation should be gentle so as to avoid damaging fragile specimens. It is also possible to slowly tumble the dishes containing the specimens. Independently of how

agitation is achieved, the buffer should flow over the specimen in order to get DNA into solution. With some buffers, longer incubation times (i.e., 5–7 days) seem to be beneficial (27), but with GuSCN buffer, a period of 1–2 days is in most cases sufficient (27). With specimens such as arthropods, incubation times of a few hours have been shown to be sufficient (28). If contamination of samples with DNA from other species (e.g., human) or other individuals of the same species (cross-contamination) is a problem, it may be beneficial to incubate samples overnight, discard (or store for potential later uses) the extraction buffer, and then incubate the samples again, and only use the extraction buffer from the second round of incubation for further processing (see Chap. 14).

6. The volume of silica suspension has to be adjusted proportionally when different extraction/binding buffer volumes are used. The volume of silica suspension used should be at least 50 μL, as smaller silica volumes yield less DNA. If very large volumes of extraction buffer are used, do not exceed 400 μL of silica suspension per extraction, as it becomes difficult to recover all of the TE used for elution. It is possible to increase the elution volume, but if more than 50 mL of extraction/binding buffer are used, purification in multiple tubes is required, although this will result in higher volumes of extract and thereby less concentrated DNA. Alternatively, when using nonchaotropic extraction buffers, it is possible to concentrate the extraction buffer prior to the adsorption step using appropriate filter systems (e.g. (32), see also Note 2).
7. Vortex silica until it is a homogenous suspension immediately before adding it to the extraction/binding buffer; note that silica particles settle relatively quickly.
8. If possible, keep the supernatant until satisfying results are obtained. If none of the samples yielded amplifiable DNA, it is possible to repeat the silica purification steps by adding freshly made silica suspension and continue from the 3-h incubation step.
9. When using the GuSCN buffer described here, a column method for washing the silica and DNA elution can be used ((31), see also Chap. 3) instead of a silica-batch extraction method. Although the efficiency of the two methods with this protocol has not been evaluated, given that they perform similarly in ancient DNA extraction (31), it is unlikely that any significant differences should occur when combined with the preceding steps of this protocol. It is also possible to use ethanol or isopropanol precipitation in combination with the GuSCN (26) or a modified (28) buffer. It should be noted that, when a nonchaotropic extraction buffer is used in combination with silica extraction, it is necessary for binding of DNA to the

silica to add a chaotropic binding buffer. In our experience, GuSCN in the concentration and volume ratio described in (31) gives the best results with regard to DNA yield and absence of PCR inhibitors.

10. If the silica is still colored after two washing steps, repeat the procedure from step 3. Washing buffer 1 normally reduces the amount of potentially inhibiting coextracted contaminants.
11. It is crucial that washing buffer 2 is removed as completely as possible at this step, as remaining traces of GuSCN can result in incomplete elution of DNA. A second elution will recover most of the remaining DNA, but this may result in lower concentrations of DNA in the final extract if pooled with the first elution (see also Note 6).
12. If more than 100 μL of silica are used for purification, it is recommended to increase the volume of the elution buffer, although the exact amount has yet to be determined experimentally. The volume of recovered elution buffer should be at least 50 μL (see also Note 6).
13. Low retention or siliconized tubes are recommended for DNA storage, as they reduce DNA loss due to tube wall effects.

Acknowledgments

I thank Beth Shapiro for pestering me until this chapter was written and the University of York for financial support.

References

1. Higuchi R, Bowman B, Freiberger M, Ryder OA, Wilson AC (1984) DNA sequences from the quagga, an extinct member of the horse family. Nature 312:282–284
2. Thomas RH, Schaffner W, Wilson AC, Pääbo S (1989) DNA phylogeny of the extinct marsupial wolf. Nature 340:465–467
3. Krajewski C, Driskell AC, Baverstock PR, Braun MJ (1992) Phylogenetic relationships of the thylacine (Mammalia: Thylacinidae) among dasyuroid marsupials: evidence from cytochrome b DNA sequences. Proc Biol Sci 250:19–27
4. Krajewski C, Buckley L, Westerman M (1997) DNA phylogeny of the marsupial wolf resolved. Proc Biol Sci 264:911–917
5. Shapiro N, Sibthorpe D, Rambaut A, Austin J, Wragg GM, Bininda-Emonds OR, Lee PL, Cooper A (2002) Flight of the dodo. Science 295:1683
6. Asher RJ, Hofreiter M (2006) Tenrec phylogeny and the noninvasive extraction of nuclear DNA. Syst Biol 55:181–194
7. Fleischer RC, James HF, Olson SL (2008) Convergent evolution of Hawaiian and Australo-Pacific honeyeaters from distant songbird ancestors. Curr Biol 18:1927–1931
8. Asher RJ, Maree S, Bronner G, Bennett NC, Bloomer P, Czechowski P, Meyer M, Hofreiter M (2010) A phylogenetic estimate for golden moles (Mammalia, Afrotheria, Chrysochloridae). BMC Evol Biol 10:69
9. Thomas WK, Pääbo S, Villablanca FX, Wilson AC (1990) Spatial and temporal continuity of kangaroo rat populations shown by sequencing mitochondrial DNA from museum specimens. J Mol Evol 31:101–112
10. Godoy JA, Negro JJ, Hiraldo F, Donázar JA (2004) Phylogeography, genetic structure and diversity in the endangered bearded vulture

(*Gypaetus barbatus*, L) as revealed by mitochondrial DNA. Mol Ecol 13:371–390

11. Leonard JA, Rohland N, Glaberman S, Fleischer RC, Caccone A, Hofreiter M (2005) A rapid loss of stripes: the evolutionary history of the extinct quagga. Biol Lett 1:291–5
12. Rohland N, Pollack JL, Nagel D, Beauval C, Airvaux J, Pääbo S, Hofreiter M (2005) The population history of extant and extinct hyenas. Mol Biol Evol 22:2435–2443
13. Krystufek B, Buzan EV, Hutchinson WF, Hänfling B (2007) Phylogeography of the rare Balkan endemic Martino's vole, *Dinaromys bogdanovi*, reveals strong differentiation within the western Balkan Peninsula. Mol Ecol 16:1221–1232
14. Moodley Y, Bruford MW (2007) Molecular biogeography: towards an integrated framework for conserving pan-African biodiversity. PLoS One 2:e454
15. Groombridge JJ, Jones CG, Bruford MW, Nichols RA (2003) 'Ghost' alleles of the Mauritius kestrel. Nature 403:616
16. Miller CR, Waits LP (2003) The history of effective population size and genetic diversity in the Yellowstone grizzly (*Ursus arctos*): implications for conservation. Proc Natl Acad Sci U S A 100:4334–4339
17. Pergams OR, Barnes WM, Nyberg D (2003) Mammalian microevolution: rapid change in mouse mitochondrial DNA. Nature 423:397
18. Miller CR, Waits LP, Joyce P (2006) Phylogeography and mitochondrial diversity of extirpated brown bear (*Ursus arctos*) populations in the contiguous United States and Mexico. Mol Ecol 15:4477–4485
19. Nyström V, Angerbjörn A, Dalen L (2006) Genetic consequences of a demographic bottleneck in the Scandinavian arctic fox. Oikos 114:84–94
20. Pergams OR, Lacy RC (2008) Rapid morphological and genetic change in Chicago-area *Peromyscus*. Mol Ecol 17:450–463
21. Endicott P, Gilbert MT, Stringer C, Lalueza-Fox C, Willerslev E, Hansen AJ, Cooper A (2003) The genetic origins of the Andaman Islanders. Am J Hum Genet 72:178–184
22. Vigilant L, Hofreiter M, Siedel HA, Boesch C (2001) Paternity and relatedness in wild chimpanzee communities. Proc Natl Acad Sci USA 98:12890–12895
23. Wandeler P, Hoeck PEA, Keller LF (2007) Back to the future: museum specimens in population genetics. Trends Ecol Evol 22:634–642
24. Miller W, Drautz DI, Janecka JE, Lesk AM, Ratan A, Tomsho LP, Packard M, Zhang Y, McClellan LR, Qi J, Zhao F, Gilbert MT, Dalén L, Arsuaga JL, Ericson PG, Huson DH, Helgen KM, Murphy WJ, Götherström A, Schuster SC (2009) The mitochondrial genome sequence of the Tasmanian tiger (*Thylacinus cynocephalus*). Genome Res 19:213–220
25. Mundy NI, Unitt P, Woodruff DS (1997) Skin from feet of museum specimens as a non-destructive source of DNA for avian genotyping. Auk 114:126–129
26. Wisely SM, Maldonado JE, Fleischer RC (2004) A technique for sampling ancient DNA that minimizes damage to museum specimens. Conserv Genet 5:105–107
27. Rohland N, Siedel HA, Hofreiter M (2004) Nondestructive DNA extraction method for mitochondrial DNA analyses of museum specimens. Biotechniques 36(814–6):818–821
28. Rowley DL, Coddington JA, Gates MW, Norrbom AL, Ochoa RA, Vandenberg NJ, Greenstone MH (2007) Vouchering DNA-barcoded specimens: test of a nondestructive extraction protocol for terrestrial arthropods. Mol Ecol Notes 7:915–924
29. Gilbert MT, Moore W, Melchior L, Worobey M (2007) DNA extraction from dry museum beetles without conferring external morphological damage. PLoS One 2:e272
30. Thomsen PF, Elias S, Gilbert MT, Haile J, Munch K, Kuzmina S, Froese DG, Sher A, Holdaway RN, Willerslev E (2009) Non-destructive sampling of ancient insect DNA. PLoS One 4:e5048
31. Rohland N, Siedel H, Hofreiter M (2010) A rapid column-based ancient DNA extraction method for increased sample throughput. Mol Ecol Resour 10:677–683
32. Noonan JP, Hofreiter M, Smith D, Priest JR, Rohland N, Rabeder G, Krause J, Detter JC, Pääbo S, Rubin EM (2005) Genomic sequencing of Pleistocene cave bears. Science 309: 597–599

Chapter 14

Case Study: Using a Nondestructive DNA Extraction Method to Generate mtDNA Sequences from Historical Chimpanzee Specimens*

Elmira Mohandesan, Stefan Prost, and Michael Hofreiter

Abstract

A major challenge for ancient DNA (aDNA) studies using museum specimens is that sampling procedures usually involve at least the partial destruction of each specimen used, such as the removal of skin, pieces of bone, or a tooth. Recently, a nondestructive DNA extraction method was developed for the extraction of amplifiable DNA fragments from museum specimens without appreciable damage to the specimen. Here, we examine the utility of this method by attempting DNA extractions from historic (older than 70 years) chimpanzee specimens. Using this method, we PCR-amplified part of the mitochondrial HVR-I region from 65% (56/86) of the specimens from which we attempted DNA extraction. However, we found a high incidence of multiple sequences in individual samples, suggesting substantial cross-contamination among samples, most likely originating from storage and handling in the museums. Consequently, reproducible sequences could be reconstructed from only 79% (44/56) of the successfully extracted samples, even after multiple extractions and amplifications. This resulted in an overall success rate of just over half (44/86 of samples, or 51% success), from which 39 distinct HVR-I haplotypes were recovered. We found a high incidence of C to T changes, arguing for both low concentrations of and substantial damage to the endogenous DNA. This chapter highlights both the potential and the limitations of nondestructive DNA extraction from museum specimens.

Key words: Ancient DNA, Chimpanzees, DNA damage, Genetic diversity, Mitochondrial DNA (mtDNA), Museum collections, Non-destructive DNA extraction, Phylogeography, Population extinction

* *Note*: In the case study presented in this chapter, we describe DNA extraction and amplification of mitochondrial DNA from historic chimpanzee samples from museum collections using a method similar to that presented in Chapter 13. We discuss specific challenges associated with nondestructive DNA extraction, including contamination and DNA damage.

Beth Shapiro and Michael Hofreiter (eds.), *Ancient DNA: Methods and Protocols*, Methods in Molecular Biology, vol. 840, DOI 10.1007/978-1-61779-516-9_14, © Springer Science+Business Media, LLC 2012

1. Introduction

Museum specimens represent one of the major sources of ancient DNA. Museum collections are valuable because they often contain rare or extinct species as well as large numbers of conspecific specimens that can be used to reveal the biological history of species and populations. Methods for DNA extraction from bones, teeth, and skin are well established (1, 2). However, for almost all of these, a piece of tooth, bone, or skin has to be removed and dissolved prior to DNA extraction.

To circumvent this limitation, a nondestructive DNA extraction method has been developed, with a reported success rate of 90% for bones up to 164 years old (3). The protocol, described in detail in Chapter 13, involves soaking the sample in GuSCN buffer and subsequently processing the buffer. Because it does not require the removal of a large piece of the specimen, this method prevents significant damage to the specimen, leaving it intact for future analyses. In addition, if necessary, the DNA extraction can be repeated 3–5 times without significant damage occurring to the specimen (3).

Here, we apply this nondestructive DNA extraction method to a large number of museum-preserved chimpanzee specimens. We discuss the success rate of this method, problems that arise during the procedure, and phylogenetic analyses performed subsequent to extraction and sequencing.

Common chimpanzees (*Pan troglodytes*) are traditionally divided into three populations or subspecies based on geographic barriers (mostly rivers): west African *P. t. verus* (4), central African *P.t. troglodytes*, and east African *P. t. schweinfurthii* (5, 6). Additional sampling in northern Cameroon/southern Nigeria has led to the designation of a fourth chimpanzee subspecies, *P. t. vellerosus* (7–11), although the phylogenetic distinctiveness and therefore the validity of this fourth chimpanzee subspecies is still debated (12). A recent analysis of about 300 microsatellites demonstrated convincingly that low levels of gene flow are occurring among the three traditionally accepted chimpanzee subspecies (12). However, due to a lack of captive individuals of *P. t. vellerosus*, the status of this potential subspecies has yet to be ascertained (12). Because chimpanzee populations have declined severely during the last decades (13–15), accessing genetic material from historic chimpanzee specimens should allow a better understanding of the geographical distribution and the population history of chimpanzees.

2. Materials and Methods

2.1. Sample Preparation

We used two rooms during the experiment so that sample preparation could be kept separate from contamination-susceptible steps including buffer preparation and PCR setup. In the second room, we carried out buffer preparation and setup of PCR reagent mix in one fume hood, and DNA extraction and the addition of DNA extract to the PCR in a second hood. In order to prevent modern DNA from potentially contaminating the experiments, we washed all working surfaces with 10–13% sodium hypochlorite solution (bleach) prior to DNA extraction. Both rooms were designated for ancient DNA work, and were spatially separated from all laboratories in which work on modern DNA was performed. The ancient DNA clean rooms were further isolated from any other area by an ante-room, which was used for decontaminating consumables and changing clothes.

We collected teeth from 86 chimpanzee (*Pan troglodytes*) individuals originating from different geographical locations in Africa and that are currently held in different museum collections. The final data set comprised specimens from 35 eastern, 20 central, two western and one western/central (Nigeria-Cameroon) locations.

2.2. DNA Extraction and Amplification

Prior to extraction, we prepared TE buffer, extraction solution, binding buffer, washing buffer, and silica suspension as described in Chapter 13. We designed two overlapping primer pairs (A and B; see Table 1) using Primer 3 version 0.4.0 (http://frodo.wi.mit.edu/primer3/). The primers were synthesized in 100 μM stock concentration and stored at −20°C. For use in PCR, we diluted the primers to 10 μM concentration with HPLC-grade water and stored them at −20°C.

Table 1
Primer designed for amplifying the investigated D-loop region of chimpanzee mtDNA

	Primer sequence 5–3′	Product size
Primer pair A		
Outer sense2 (OS2)	5′-CGC TAT GTA TTT CGT ACA TTA CT-3′	210 bp
Inner antisense3 (IAS3)	5′-RTA GGT TTG TTG ATA TYR G-3′	
Primer pair B		
Inner sense3 (IS3)	5′-TCA ACT CTC AAC TRT CRM ACA TA-3′	130 bp
Outer antisense2 (OAS2)	5′-GAT TTG ACT GTA ATG TGC TAT G-3′	

For extraction, we first cleaned the surface of each specimen using a tissue moistened with HPLC-grade water. Removing dirt from the surface of the samples reduces the amount of substances that might inhibit the DNA extraction and/or the following enzymatic manipulations of DNA extract such as PCR.

We then soaked the samples in 5 mL extraction solution (L6 buffer) and incubated them at room temperature in the dark with constant slow rotation. After 5–7 days, we removed the buffer and rinsed the sample with HPLC-grade water. We dried the samples at room temperature in preparation for return to the museums from which they were obtained.

To continue with the DNA extraction, we transferred the buffer into a new 15-mL centrifuge tube. We added 50–100 μL of silica suspension (after vortexing the silica suspension to be certain that it was adequately mixed) and incubated the mixture for 1–3 h at room temperature with rotation. We then centrifuged the buffer at 1,800 × *g* for 2 min and either discarded the supernatant or stored it at 4°C for later use. Next, we washed the silica pellet with 1 mL L2 buffer by pipetting up and down. We transferred the resuspended mixture to a 2-mL Eppendorf tube. This transfer makes handling more convenient, as 2-mL tubes rather than 15-mL tubes can be used in all the following steps. We pelleted the silica via centrifugation for 5 s at 16,000 × *g*, discarded the supernatant, and carefully removed any remaining liquid using a 200-μL pipette. If the binding solution (L2 buffer) is not completely removed in this step, the salt concentration in the elution buffer will be too high, thus preventing the DNA from being completely released from the silica during elution.

We then washed the pellet with 1 mL washing buffer by pipetting up and down. We centrifuged the resuspended mixture for 10 s at 16,000 × *g*. We discarded the supernatant and removed the remaining liquid again carefully with a pipette. We dried the pellets at 56°C for 5 min or approximately 15 min at room temperature with open lids. We then added 100 μL TE (1×) to the pellet, incubated the mixture for 8 min at 65°C, and resuspended the pellet by stirring with the pipette tip and pipetting up and down. Finally, we centrifuged the eluate at 16,000 × *g* for 1 min and transferred the supernatant into a new 2-mL Eppendorf tube, being careful not to leave any trace of silica. For some specimens, second and third extractions starting at the incubation step were subsequently performed (see Subheading 3).

We used the obtained extracts to generate an approximately 225 bp fragment of the HVR-I region of chimpanzee mtDNA by PCR amplifying two overlapping fragments of 210 and 130 bp, respectively, using primer pairs A (OS2/IAS3) and B (IS3/OAS2; see Table 1). PCR was carried out in 20 μL volumes containing 1× PCR buffer (Applied Biosystems), 4 mM $MgCl_2$ (Applied Biosystems), 1 mg/mL BSA (Invitrogen), 0.5 mM mixed dNTPs (in equal concentrations; Amersham Biosciences), 0.25 μM of each primer (MWG-Biotech AG), 0.5–1 U of Taq Gold DNA polymerase (Applied Biosystems), and 5 μL DNA template (irrespective of DNA concentration). The initial denaturation step (94°C for 4 min) was followed by 60 cycles of denaturation at 93°C for 20 s, binding of primers at 51°C (primer pair A) and 53°C (primer pair B) for 30 s and strand replication at 72°C for 30 s, followed by a final extension at 72°C for 10 min. The PCR products were subjected to electrophoresis in 1.5% agarose, stained with ethidium bromide (50 ng/mL) and visualized over UV light. We included one negative control for every seven PCR reactions. Each fragment was amplified twice for each specimen.

We purified PCR products of the expected length with the QIAquick Gel Extraction Kit (QIAGEN, Germany), and cloned them using the TOPO TA® Cloning Kit (Invitrogen, The Netherlands) according to the manufacturer's instruction. We sequenced the insert sequences for eight clones per sample on an ABI 3700 capillary sequencer after colony PCR and purification on a QIAGEN BioRobot 9600.

2.3. Phylogenetic Analysis

We aligned the nucleotide sequences from the HVR-I regions sequenced from 56 chimpanzees in BioEdit version 7.0 (16) using CLUSTAL-W software. We checked the authenticity of obtained DNA sequences using BlastSearch (National Center for Biotechnology Information) (17) and reconstructed the phylogenetic relationship between the recovered sequences as well as extant chimpanzee sequences obtained from GenBank by constructing a serial network (18). The serial network was created using the open-source R script TempNet (available at www.stanford.edu/group/hadlylab/tempnet/). TempNet uses statistical parsimony to illustrate within-species relationships through time.

3. Results and Discussion

Using the silica-based nondestructive method, we successfully amplified and sequenced mtDNA sequences from 65% (56 of 86) of the chimpanzee specimens that were stored in different museums. Of these, 53 samples (95%) yielded both PCR products, while the remaining three samples (5%) could only be partially amplified.

All recovered sequences showed between 98 and 100% BLAST similarity to chimpanzee mtDNA sequences archived in GenBank. Analysis of consensus and clone sequences generated from two independent PCRs revealed identical sequences for 29 museum specimens (apart from C to T changes in individual clones, which are almost certainly due to DNA damage; see below) and multiple sequence variants within the remaining 26 (one sample could only be amplified once and was excluded from further analyses). Thus, just over half of the samples yielded identical sequences across multiple PCRs, although for six of the samples yielding additional sequences, these occurred at such a low frequency that a likely endogenous sequence could be inferred. This overall result most likely indicates that cross-contamination occurred between museums specimens, especially since the sequence variants recovered sometimes belong to different chimpanzee subspecies. To investigate this further, we performed additional nondestructive extractions on 16 of the specimens that had yielded ambiguous sequences. This additional experiment was motivated by the realization that the first extraction may recover not only endogenous DNA but also any potential surface contaminant DNA, including cross-contamination that may have occurred as researchers handled multiple specimens. Additional extractions performed after the first extraction should therefore be less likely to recover surface contaminants.

We performed second and in some cases third DNA extractions from 16 of the samples with variant sequences. Each extraction yielded less amplifiable DNA than the previous extraction, as judged by the number of failing PCRs and the strength of the product when amplifications were successful. However, the amount of DNA contamination was also reduced to some extent, and a likely endogenous DNA sequence could eventually be deduced for 9 of these 16 samples, while the remaining seven samples could not be resolved. Thus, in total we were able to recover reproducible sequences from 44 samples, resulting in a total of 39 distinct haplotypes.

This result is in stark contrast to previous experience with this protocol when no evidence for contamination was observed (3, 19, 20). However, while it should be noted that two of these previous studies were performed on small mammal specimens, where both storage conditions and, due to the fragile nature of the specimens, extraction kinetics might be different, the initial study introducing this method used both chimpanzee and hyena teeth. It is not clear why the results of this study differ so much from those of previous studies. One potential cause may lie in differences in museum storage and handling conditions that might have facilitated cross-contamination among the samples used in this study, but it is impossible to ascertain this possibility. Another fact worth mentioning is a high incidence of C to T changes, indicative of DNA damage (21) in our results. Thus, of the 29 samples that yielded unambiguous sequences, 26 showed C to T changes in individual clones. This observation suggests not only high DNA damage but also low DNA concentrations in these samples, making them more susceptible to contamination. Independent of the eventual cause for the high contamination rate on the samples used, our results show that studies on museum specimens face similar problems as those using fossil DNA, at least when using this extraction method. Therefore, similar precautions such as multiple

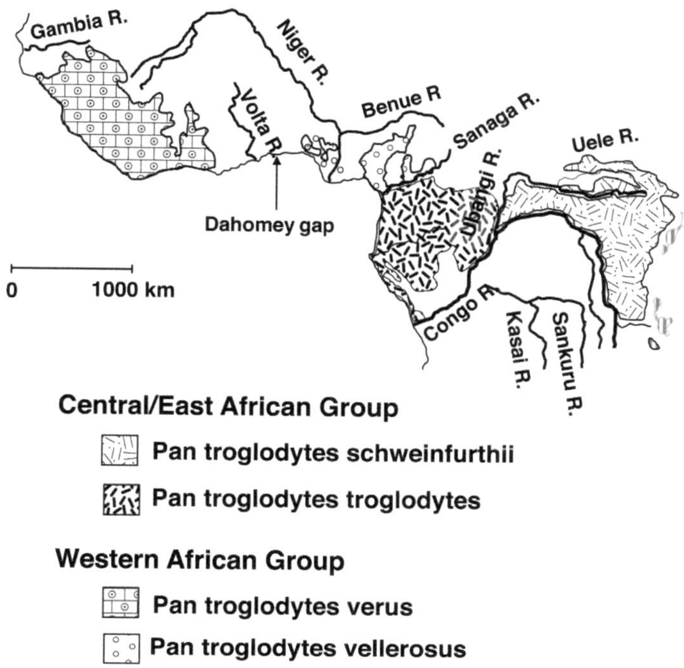

Fig. 1. Geographical distribution of chimpanzee subspecies.

extractions and amplifications as well as obtaining multiple clonal sequences are an absolute requirement in such studies.

Chimpanzee subspecies are divided into two geographically and genetically defined groups: a central/eastern African group (*P. t. schweinfurthii* and *P. t. troglodytes*) and a western African group (*P. t. verus* and *P. t. vellerosus*) with a significant phylogeographic break at the Sanaga River in central Cameroon (Fig. 1). A temporal network (18) reconstructed from our historical sequences and modern chimpanzee sequences obtained from GenBank shows that all historical haplotypes are closely related to modern ones (Fig. 2), although some of them have not (yet) been found in the extant gene pool.

With 51%, the DNA extraction success rate in this study is lower than in previous studies reporting the method (3, 19, 20), but still sufficiently high to obtain DNA from about half of the investigated specimens. Similarly, the length of the obtained PCR products is large enough to obtain, by using several overlapping fragments, DNA sequences sufficiently long for phylogeographic and phylogenetic analyses. However, the high incidence of contaminating sequences found also indicates that a substantial failure rate has to be taken into account when planning a study, although there seem to be large differences among collections and species, probably depending on storage and handling.

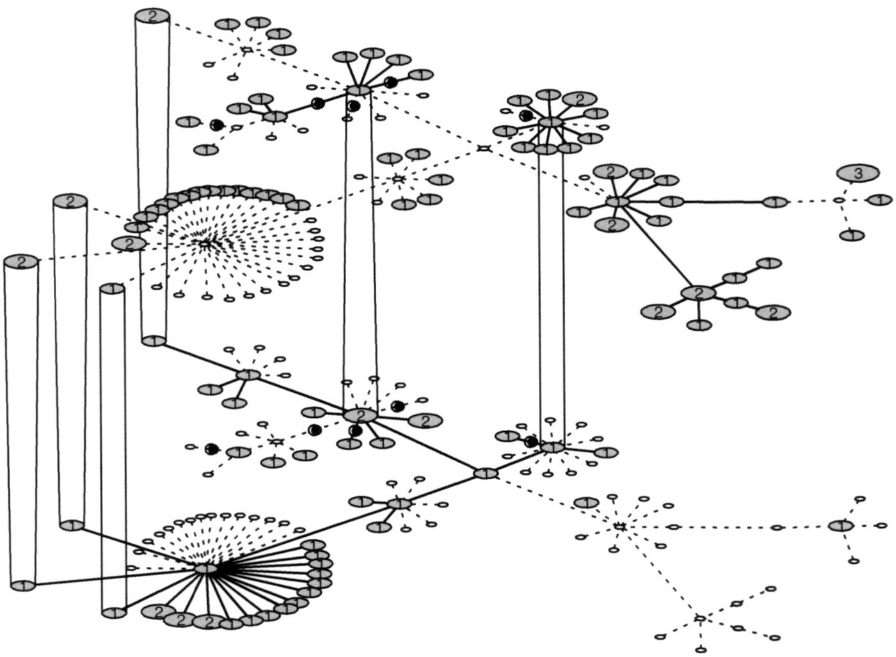

Fig. 2. Temporal statistical parsimony network of modern and ancient chimp sequences. The upper layer comprises modern-day sequences obtained from GenBank, whereas the lower layer consists of ancient DNA samples generated in this study. Haplotypes sampled in a given time layer are represented as *gray ellipses*. Those present in the overall network, but not in the individual time layer are shown as *small white ellipses*. Haplotypes shared between the two layers are connected by *vertical lines*. Haplotypes present in a time-horizon are connected by *solid lines*, whereas lines connecting at least one unsampled haplotype for this time-horizon are *dotted*. Those separated by more than one mutation are indicated by one *small black circle* for each additional mutation. Please note that for graphical reasons, not all modern sequences available were used in the network. Therefore, a larger proportion of museum sequences than shown in this figure are actually still present in the modern population.

Both success rate and total length of the DNA sequences that can be obtained should increase considerably when using DNA hybridization capture methods (22–25) rather than PCR for targeting specific DNA regions. These methods have recently been used successfully for targeting both mitochondrial (up to complete mtDNA genomes (25, 26)) and nuclear DNA (27). Due their ability to target very short DNA fragments, they are ideally suited for the analysis of fragmented DNA such as that recovered from museum specimens. It needs to be noted, though, that measures used to distinguish endogenous ancient DNA obtained from Pleistocene specimens from contaminating modern DNA such as fragmentation or nucleotide substitution patterns may not be applicable to museum specimen DNA for several reasons. First, due to their younger age, museum specimens may not have accumulated DNA damage to the extent that fossil DNA dating to the Pleistocene has. Second, and perhaps even more importantly, the sequences contaminating museum specimens probably originate quite frequently from cross-contamination with DNA from other museum specimens, which is likely to display highly similar damage patterns. However, as our results show, this problem can be addressed at least partially by performing two consecutive extractions and by preferential use of the second extract.

Acknowledgments

We thank the lab members of the Research Group Molecular Ecology at the Max Planck Institute for Evolutionary Anthropology, and especially Tim Heupink, for their assistance in laboratory work, museum curators for providing us with chimpanzee samples, and the Max Planck Society for financial support.

References

1. Krings M, Stone A, Schmitz RW et al (1997) Neandertal DNA sequences and the origin of modern humans. Cell 90:19–30
2. Hadly EA, Kohn MH, Leonard JA et al (1998) A genetic record of population isolation in pocket gophers during Holocene climatic change. Proc Natl Acad Sci USA 95:6893–6896
3. Rohland N, Siedel H, Hofreiter M (2004) Nondestructive DNA extraction method for mitochondrial DNA analyses of museum specimens. Biotechniques 36(814–816):818–821
4. Schwarz E (1934) On the local races of chimpanzees. Ann Mag Nat Hist Lond 13: 576–583
5. Hill WCO (1969) In: Bourne GH (ed) The chimpanzee; a series of volumes on the chimpanzee, vol. 1. S. Karger AG, Basel, NY, pp. 22–49
6. Groves CP (2001) Primate taxonomy. Smithsonian Institution Press, Washington, DC, 350p
7. Gonder MK, Oates JF, Disotell TR et al (1997) A new West African chimpanzee subspecies? Nature 388:337
8. Gonder MK, Disotell TR, Oates JF (2006) New genetic evidence on the evolution of chimpanzee populations and implications for taxonomy. Int J Primatol 27:1103–1127
9. Gonder MK (2000) Evolutionary genetics of chimpanzees (*Pan troglodytes*) in Nigeria and Cameroon. Ph.D. Dissertation, City University of New York, New York, 338pp
10. Gonder MK, Disotell TR (2006) Contrasting phylogeographic histories of chimpanzees in Nigeria and Cameroon: a multilocus analysis. In: Lehman S, Fleagle J (eds) Primate biogeography. Springer, New York, pp 129–161
11. Gonder MK, Disotell T, Oates JF (2006) New genetic evidence on the evolution of chimpanzee populations and implications for taxonomy. Int J Primatol 27:1103–1127
12. Becquet C, Patterson N, Stone AC et al (2007) Genetic structure of chimpanzee populations. PLoS Genet 3:e66
13. Campbell G, Kuehl H, N'Goran KP et al (2008) Alarming decline of West African chimpanzees in Côte d'Ivoire. Curr Biol 18:903–904
14. Walsh PD, Abernethy KA, Bermejo M et al (2003) Catastrophic ape decline in western equatorial Africa. Nature 422:611–614
15. Greengrass E (2009) Chimpanzees are close to extinction in Southwest Nigeria. Prim Cons 24:77–83
16. Hall TA (1999) BioEdit: a user-friendly biological sequence alignment editor and analysis program for Windows 95/98/NT. Nucl Acids Symp Ser 41:95–98
17. Altschul SF, Madden TL, Schäffer AA et al (1997) Gapped BLAST and PSI-BLAST: a new generation of protein database search programs (Review). Nucleic Acids Res 25: 3389–3402
18. Prost S, Anderson CNK (2011) TempNet: a method to display statistical parsimony networks for heterochronous DNA sequence data. Methods Ecol Evol 2:663–667. doi:10.1111/j.2041-210X.2011.00129.x

19. Asher RJ, Hofreiter M (2006) Tenrec phylogeny and the noninvasive extraction of nuclear DNA. Syst Biol 55(2):181–194
20. Asher RJ, Maree S, Bronner G, Bennett NC, Bloomer P, Czechowski P, Meyer M, Hofreiter M (2010) A phylogenetic estimate for golden moles (Mammalia, Afrotheria, Chrysochloridae). BMC Evol Biol 10:69
21. Hofreiter M, Jaenicke V, Serre D, Haeseler Av A, Pääbo S (2001) DNA sequences from multiple amplifications reveal artifacts induced by cytosine deamination in ancient DNA. Nucleic Acids Res 29(23):4793–4799
22. Hodges E, Xuan Z, Balija V et al (2007) Genome-wide in situ exon capture for selective resequencing. Nat Genet 39:1522–1527
23. Hodges E, Rooks M, Xuan Z et al (2009) Hybrid selection of discrete genomic intervals on custom-designed microarrays for massively parallel sequencing. Nat Protoc 4:960–974
24. Gnirke A, Melnikov A, Maguire J et al (2009) Solution hybrid selection with ultra-long oligonucleotides for massively parallel targeted sequencing. Nat Biotechnol 27:182–189
25. Briggs AW, Good JM, Green RE et al (2009) Targeted retrieval and analysis of five Neandertal mtDNA genomes. Science 325:318–321
26. Krause J, Briggs AW, Kircher M et al (2010) A complete mtDNA genome of an early modern human from Kostenki, Russia. Curr Biol 20:231–236
27. Burbano HA, Hodges E, Green RE et al (2010) Targeted investigation of the Neandertal genome by array-based sequence capture. Science 328:723–725

Chapter 15

PCR Amplification, Cloning, and Sequencing of Ancient DNA

Tara L. Fulton and Mathias Stiller

Abstract

PCR amplification of DNA is routine in modern molecular biology. However, the application of PCR to ancient DNA (aDNA) experiments often requires significant modification to standard protocols. The degraded nature of most aDNA fragments requires targeting shorter fragments, performing replicate amplifications, incorporating multiple negative controls, combating PCR inhibition, using specific DNA polymerases to deal with damaged bases, working in a separate aDNA facility, and modifying the PCR recipe to deal with damaged and low copy-number target DNA. In this chapter, we describe how and why these procedures are implemented, discuss aDNA-specific troubleshooting methodology, and suggest modifications to commercial cloning and sequencing procedures to reduce the expense of PCR product cloning.

Key words: Polymerase chain reaction, PCR optimization, BSA, inhibition, Ancient DNA, DNA polymerase

1. Introduction

The invention of the polymerase chain reaction (PCR) (1) revolutionized the field of ancient DNA (aDNA) research. In theory, only a single copy of the targeted DNA region is required for PCR, making it a powerful tool for amplifying aDNA from samples where only a handful of intact copies of the target region may remain. PCR is not, by any means, a technique exclusive to aDNA research. However, its use with aDNA requires modifications to the experimental design, the experiment itself, and post-experimental troubleshooting.

Ancient DNA is often highly degraded, and even exceptionally preserved permafrost specimens may contain only 5% of surviving DNA fragments longer than 300 base pairs (bp) (2). Thus, when fragments longer than 100–300 bp are targeted using PCR, it is

Beth Shapiro and Michael Hofreiter (eds.), *Ancient DNA: Methods and Protocols*, Methods in Molecular Biology, vol. 840, DOI 10.1007/978-1-61779-516-9_15, © Springer Science+Business Media, LLC 2012

possible that long fragments of undamaged, modern DNA may be preferentially amplified. To overcome this, a series of overlapping primer sets can be used to obtain a long stretch of continuous DNA sequence in small, stepwise fragments. This has the added advantage of identifying any nontarget amplifications such as numts (nuclear insertions of the mitochondrial DNA), other pseudogenes, or nonhomologous copies of the target gene, if mismatches are observed between overlapping regions of the amplified fragments. It is also routine to clone at least some of the amplification products of aDNA experiments, as this can identify potential contaminants or PCR artifacts and allow evaluation of the extent of post-mortem damage.

The high-performing Platinum Taq High Fidelity and AmpliTaq Gold (both from Life Technologies) are among the most common polymerases used in aDNA experiments. The choice of polymerase is important as commercial polymerases vary widely in their efficiency in synthesizing aDNA (3) and in the particular way they interact with damaged bases (4). Even with high-fidelity polymerases, it is important to consider the possibility of strand jumping which can produce chimeric products. An additional benefit of both Platinum Taq and AmpliTaq Gold is that they are hot-start polymerases, a desirable attribute as PCR amplification from ancient extracts is generally set up in a facility that is spatially distant from the thermocycler.

PCR inhibitors are often co-extracted with aDNA, as samples have often been exposed to environmental contaminants for tens of thousands of years. To minimize inhibition, serum albumin, and commonly bovine serum albumin (BSA), can be included in aDNA PCR. BSA binds PCR-inhibiting co-extracts and prevents target DNA from adhering to the tube rather than being amplified. Including BSA can dramatically improve PCR success (3) and is useful as a troubleshooting measure when PCRs are unsuccessful.

DNA damage is also common in aDNA extracts. Several measures have been recommended to deal with damage in PCR of ancient specimens, including pretreatment with uracil DNA glycosylase (UNG or UDG) to remove uracil (5) or *N*-phenacylthiazolium bromide (PTB) to cleave crosslinks (6). Although treatments designed to remove uracil can be beneficial, many aDNA researchers are reassured of the authenticity of the resulting ancient sequences when random C–T (or G–A, if the cloned product is the reverse strand pairing to the strand on which the damage occurred) transitions are observed in cloned products of a PCR, as this form of damage is common in ancient samples.

It is important to note that aDNA PCR will often require much more optimization than modern DNA PCR because template quantity, quality, and level of inhibition are unique to each extract. The quantity and quality of starting template copies can be highly stochastic even between aliquots of a single DNA extract, so multiple PCRs from different extracts are suggested to evaluate the consistency of sequencing results and confirm the consensus sequence. Cloning multiple PCR amplifications derived from poorly preserved samples is highly recommended. Cloning only a single PCR product may be misleading if the starting copy is damaged, and a miscoded base is incorporated in an early PCR cycle.

As this book is targeted for specifically aDNA research, familiarity with basic PCR and routine molecular biology lab protocols, such as running an agarose gel and pipetting, is assumed. We refer readers with no experience with PCR or basic molecular biology methods to more general works (7, 8).

2. Methods

All reagents and plastics should be sterile, DNA and DNAse free. All solutions should be molecular biology grade or similar.

2.1. Required Reagents

1. Deoxynucleoside triphosphates (dNTPs) of 100 μM each, combined in equal volume to yield a dNTP mix of 25 μM each dATP, dGTP, dTTP, and dCTP.
2. DNA Polymerase + buffer supplied with polymerase (see Note 1).
3. Magnesium ions supplied separately with polymerase usually as $MgCl_2$ or $MgSO_4$.
4. Forward and reverse primers diluted to 10 μM each.
5. BSA, rabbit serum albumin (RSA), or a different serum albumin prepared to 10 mg/mL solution in sterile water (see Note 2).
6. Barrier/filter tips and PCR reaction tubes/plates.
7. Thermocycler with heated lid.
8. DNA template.

2.2. Agarose Gel Visualization

1. 2% agarose gel.
2. 50× TAE (500 mL: 121 g Tris, 28.6 mL glacial acetic acid, 50 mL 0.5 M EDTA pH 8.0), diluted to 1× for running buffer.
3. 6× loading dye (0.25% Orange-G (TCI), 0.1875% xylene cyanol (IBI Scientific), 30% glycerol).
4. DNA ladder. The ladder can be diluted with TE buffer: 125 μL (0.25 μg) prepared ladder + 1,125 μL TE + 250 μL 6× loading dye (supplied with ladder).
5. Agarose gel electrophoresis rig and power supply.
6. Ethidium bromide (EtBr) and a UV transilluminator.

2.3. PCR Purification

1. A commercial PCR purification kit, e.g., Qiagen, Millipore, ExoSAP, Agencourt AMPure XP.

2.4. BigDye (Applied Biosystems) Sequencing

1. BigDye sequencing kit (Applied Biosystems).
2. 1 μM primer.
3. A sequencing cleanup method, e.g., Ethanol–EDTA (per the *BigDye manual*), sephadex-based methods such as Qiagen DyeEx.
4. Refrigerated plate centrifuge (depends on cleanup method selected).

2.5. Cloning (Using the TOPO-TA Kit, Invitrogen)

1. TOPO-TA cloning kit (Invitrogen).
2. Agar plates containing X-gal, IPTG, and Ampicillin per the TOPO manual for blue–white screening of plasmid-containing colonies.
3. PCR reagents as listed in Subheading 2.1 available in the modern lab (hot-start *Taq* and BSA are not necessary).
4. Water bath.
5. Incubator.
6. Bunsen burner.
7. Fumehood (recommended for handling bacteria, but not required).

3. Methods

While setting up the reactions, open the reagent containers, tip boxes, PCR tubes, etc., *only* when pipetting in or out. This will greatly reduce any potential contamination transmitted by aerosols.

3.1. Master Mix Setup

1. As all of the PCRs will use the same basic recipe, plan out a master mix of the ingredients common to all reactions (see Note 3). Always include at least one PCR negative control (no DNA extract) reaction per 8–10 sample reactions (see Note 4). Generally, PCR positive controls are avoided in aDNA. However, if a positive control is necessary, use another ancient sample as this control (see Note 5).
   (a) 1–2 units of polymerase (<0.25 μL Platinum Taq High Fidelity or AmpliTaq Gold) (see Note 6).
   (b) 1× buffer (2.5 μL 10× PCR buffer).
   (c) 0.25–0.625 mM each dNTP (0.25–0.625 μL of 25 mM dNTPs).
   (d) 2–4 mM magnesium (i.e., 1–2 μL of $MgSO_4$ for HiFi or 2–4 μL of $MgCl_2$ for AmpliTaq Gold) (see Note 7).

(e) 1–2 μg serum albumin (2.5–5 μL of 10 mg/mL solution).

(f) Purified water to 25 μL; accounting for primers and template volume (below).

Follow the manufacturer's protocol provided with the polymerase to set up a 25 μL reaction. This will generally involve the following:

2. Mix the master mix by flicking it with a fingertip and inverting the tube several times. Dispense the master mix into all of the tubes.
3. If only one set of primer is used, add them to the reaction mix: 0.2–0.4 μM each primer (0.5–1 μL of 10 μM stock). If not, add the primers to each set of reactions separately by preparing a primer mix in the cap of the PCR tube of the negative and then dispense into the caps of the reactions that are designated to contain sample.
4. Add template to tubes individually. Ensure all tubes are closed before opening any of the template tubes (see Note 8). Dispense the template directly into the mix in the tube. Generally 0.5–1.0 μL (1–5% of the DNA extract's total volume) of template is used (see Note 9).
5. Spin down the reaction tubes briefly, and place them in the thermocycler. Use the basic cycling conditions as suggested for the polymerase manufacturer, paying attention to include an initial hot-start period, if applicable (see Note 10). Due to the generally low number of initial template molecules, increasing the number of cycles may provide greater yield. It is common to perform 40–60 PCR cycles with aDNA.

3.2. Agarose Gel Visualization and PCR Purification

1. Prepare a 2% or higher concentration agarose gel in TAE (or TBE). EtBr may either be included in the buffer, the gel, or applied as a post-stain. EtBr is a mutagen and must be handled with care (see Note 11).
2. Run out 2–5 μL of the completed PCR (see Note 12). Visualize with UV light. If bands of the expected length are present in the negative(s), repeat the PCR. If no bands are present at all, try increasing or decreasing either the stringency of the reaction or the amount of template used (see Note 9). If many bands are observed, an increase in stringency is required (use less magnesium and/or increase the annealing temperature).
3. If the PCR yields a single, clean band, purify the remaining 20 μL of the reaction by removing any unincorporated reagents via your favorite commercial PCR purification system. To obtain a more concentrated PCR product for direct sequencing, elute the reactions in less eluate than normal (~50% of normal eluate volume).

3.3. Direct Sequencing

1. Quantify the product using a nanospec or by visual estimation on agarose (compared to the ladder of known concentration).
2. Sequence using BigDye v3.1 (or the latest version) chemistry and recommended cycling conditions. This chemistry can be diluted for economy, because long reads are not necessary for the short aDNA PCR target fragments. Due to small volumes, this should be prepared as a large master mix based on a single reaction:
   (a) 0.25 μL BigDye v3.1.
   (b) 1.75 μL BigDye v3.1buffer.
   (c) 6.4 μL purified template + water (~30–60 ng template works well).
   (d) 1.6 μL 1 μM primer.
3. Purify the sequencing reaction using EDTA–Ethanol precipitation as described in the *BigDye manual* or by sephadex-based methods. Depending on the strength of the reaction, 50–100% of the sequencing reaction will need to be loaded for detection on a capillary sequencer (see Note 13).

3.4. Cloning

1. Clone PCR products using any PCR product cloning protocol. TOPO-TA cloning (Invitrogen) is quick, easy, and efficient, but expensive (see Note 14).
2. Plate out up to three plates from each cloning reaction, depending on the dilution factor of the TOPO kit. If the reaction is scaled down tenfold, plating the entire reaction on a single plate generally yields an appropriate number of colonies.
3. Fill a 96-well plate with 50 μL of water in each well. Any colonies that have grown up in the presence of ampicillin will contain the plasmid, which confers ampicillin resistance. By invoking a traditional blue–white screening protocol based on the modified *lac* operon system, colonies with plasmids lacking a PCR insert will be blue (all β-galactosidase enzyme fragments are produced and the substrate, X-gal, is cleaved and turns color) and colonies that hold the PCR insert will be white (no lacZα fragment is produced and X-gal is not cleaved). Pick 7–15 positive (white) colonies with separate pipette tips, placing each tip into the water of a separate well (see Note 15).
4. Place a micropipette onto the tip, swirl the tip in the water, and then blow out any liquid in the tip to ensure that the colony has transferred to the water. Eject the tip into biohazardous waste receptacle. Pick one negative (blue) colony per cloning reaction for easy visual comparison on the gel with the colonies that should have the short aDNA insert. Cover the plate with a sticker or lids and vortex vigorously to break up the colony and resuspend the cells. There is no need to heat the cells as they will lyse upon initiation of PCR cycling.

5. Amplify the insert using appropriate primers for the vector used (M13 for TOPO-TA) in the 12.5 μL reaction: 1× buffer, 0.125 μL regular *Taq* (i.e., Promega, Applied Biosystems, EconoTaq), 0.125 μL 25 mM dNTPs, 0.5 μL 25 mM $MgCl_2$, 1.25 μL 10 μM each primer, 5 μL of colony solution, and water up to 12.5 μL total volume.
6. Visualize and clean the PCR product as for the first PCR, but only load 2–3 μL of PCR product on the gel.
7. Sequence as above, but only use 0.5–1 μL of template and only load 30–50% or less of the reaction for capillary sequencing.

4. Notes

1. It is easier to optimize reactions when the magnesium is not included in the buffer.
2. Although BSA is not required for PCR to function, it is almost always used in aDNA PCR.
3. As polymerase is very expensive, there is no need to make extra mix unless you are doing many reactions at a time. If significant losses due to pipetting occur, add an extra reaction to the mix recipe.
4. Carrier DNA, such as lambda phage DNA or DNA from another nontarget species, can be added to the PCR negative to ensure that any contaminants that may be present are amplified.
5. If there is no choice but to use a modern sample as a PCR positive, set up all the reactions, take the tubes to the modern lab and open *only* the single tube to which the template is to be added. Be sure to include a modern positive control that will be easily identified as a contaminant (i.e., a genetically divergent sample). Modern samples should *never* be brought into the ancient lab.
6. Do not vortex or centrifuge the polymerase itself, only as part of the master mix.
7. If you are not using a hot-start *Taq*, the magnesium can be added separately with the primers in the reaction tube cap and only spun down immediately before cycling. Alternately, a small amount of sterile wax can be added to the master mix and melted, briefly, and the primer/*Taq* mix added on top of this protective barrier. Either of these will help to reduce the chance of nonspecific polymerase activity. A non hot-start *Taq* polymerase reaction should be set up on ice or in a cold block.
8. Tubes with individual caps are useful for this.

9. As aDNA extractions frequently contain many inhibitors, often *reducing* the amount of template can increase product yield as inhibitors are more diluted. This may be achieved by either reducing the amount of template in the PCR reaction or by diluting aliquots of the extract itself by a factor of 5–10.
10. In case of using AmpliTaq Gold, increasing the initial hot-start period from 10 min (as recommended by the manufacturer) to 12 min will give more consistent amplification results, especially when the reaction is starting from a very low number of initial template molecules.
11. Adding EtBr to the running buffer offers the most dilute option, but increases the potential for splashing and also leads to EtBr contamination of the gel apparatus. Addition of EtBr to the gel prior to setting also contaminates the apparatus, but keeps the EtBr more contained than in the buffer. However, if EtBr is added to the gel, it must *never* be added before heating, as EtBr will be aerosolized, which is hazardous. Staining the gel after electrophoresis requires a more concentrated solution of EtBr than would be added to the running buffer, but provides a very contained region of contamination and often produces sharper DNA band images. Always dispose of EtBr-contaminated waste following your institution's health and safety protocols.
12. Be sure that the samples are run out slowly so adequate separation occurs. As fragments are quite small, it is important to obtain clear differentiation from primer-dimers, which are often not much smaller than the desired product.
13. Do not load this much modern sequence—you will be greatly overloading the reaction, which is hard on the sequencer.
14. Although it is not recommended by the manufacturer, we have had success using reduced reaction volumes. The reaction can be scaled down up to tenfold, but the results become increasingly erratic with increased dilution, presumably as the reaction kinetics become less efficient. For ten cloning reactions from one tube of competent cells, follow manufacturer's protocol for TOPO-TA cloning, but modify the following volumes: use 3.5–5 ng of DNA in a total of 0.5 μL, plus 1 μL prepared reaction mix (1.25 μL salt, 1.25 μL vector, 7.5 μL water), and 5 μL competent cells. Recover in 50 μL SOC or LB media and plate everything on a single agar plate.
15. If the sample is known to be uncontaminated and not badly damaged, fewer (i.e., 3–4) clones are required. If the sample is thought to have high damage or low-level contamination from collection or storage, more clones are required (i.e., 12–16). With human work, even more clones (24+) may be desirable to detect potential contaminants.

References

1. Saiki RK, Scharf S, Faloona F, Mullis KB, Horn GT, Erlich HA, Arnheim N (1985) Enzymatic amplification of beta-globin genomic sequences and restriction site analysis for diagnosis of sickle-cell anemia. Science 230:1350–1354
2. Poinar HN, Schwarz C, Qi J, Shapiro B, MacPhee RDE, Buigues B, Tikhonov A, Huson DH, Tomsho LP, Auch A, Rampp M, Miller W, Schuster SC (2006) Metagenomics to paleogenomics: large-scale sequencing of mammoth DNA. Science 311:392–394
3. Rohland N, Hofreiter M (2007) Comparison and optimization of ancient DNA extraction. Biotechniques 42:343–352
4. Heyn P, Stenzel U, Briggs AW, Kircher M, Hofreiter M, Meyer M (2010) Road blocks on paleogenomes-polymerase extension profiling reveals the frequency of blocking lesions in ancient DNA. Nucleic Acids Res 38:e161
5. Willerslev E, Cooper A (2005) Ancient DNA. Proc Biol Sci 272:3–16
6. Poinar HN, Hofreiter M, Spaulding WG, Martin PS, Stankiewicz BA, Bland H, Evershed RP, Possnert G, Paabo S (1998) Molecular coproscopy: dung and diet of the extinct ground sloth *Nothrotheriops shastensis*. Science 281:402–406
7. Bartlett JMS, Stirling D (2003) PCR protocols, 2nd edn. Humana, Totowa, NJ
8. Sambrook J, Russell DW (2006) The condensed protocols from molecular cloning: a laboratory manual. Cold Spring Harbor Laboratory Press, Cold Spring Harbor, NY

Chapter 16

Quantitative Real-Time PCR in aDNA Research

Michael Bunce, Charlotte L. Oskam, and Morten E. Allentoft

Abstract

Quantitative real-time PCR (qPCR) is a technique that is widely used in the field of ancient DNA (aDNA). Quantitative PCR can be used to optimize aDNA extraction methodologies, to detect PCR inhibition, and to quantify aDNA libraries for use in high-throughput sequencing. In this chapter, we outline factors that need to be considered when developing efficient SYBR Green qPCR assays. We describe how to setup qPCR standards of known copy number and provide some useful tips regarding interpretation of qPCR data generated from aDNA templates.

Key words: qPCR, Ancient DNA, SYBR Green, PCR inhibition, qPCR standard, DNA extraction optimization, Library quantitation

1. Introduction

The invention of PCR has been a crucial technological advance in the field of ancient DNA (aDNA). However, conventional PCR methods, in which the success of the reactions is evaluated at the endpoint of thermocycling (typically following 40–50 cycles), should be considered qualitative, because the dynamic range of endpoint PCR is, at best, two orders of magnitude (1). Hence, it is generally difficult to tell whether a PCR reaction was seeded by ten or by ten million template molecules based on the intensity of amplicon staining on a gel. Real-time PCR methods have a dynamic range of greater than eight orders of magnitude, making them a powerful analytical tool.

An in-depth discussion of the theory of quantitative PCR (qPCR) is beyond the scope of this chapter, but can be found in a number of excellent reviews and book chapters (see (1–3)). In addition, we recommend the online resource at http://www.gene-quantification.info/. It should be noted that the bulk of the

Beth Shapiro and Michael Hofreiter (eds.), *Ancient DNA: Methods and Protocols*, Methods in Molecular Biology, vol. 840, DOI 10.1007/978-1-61779-516-9_16,

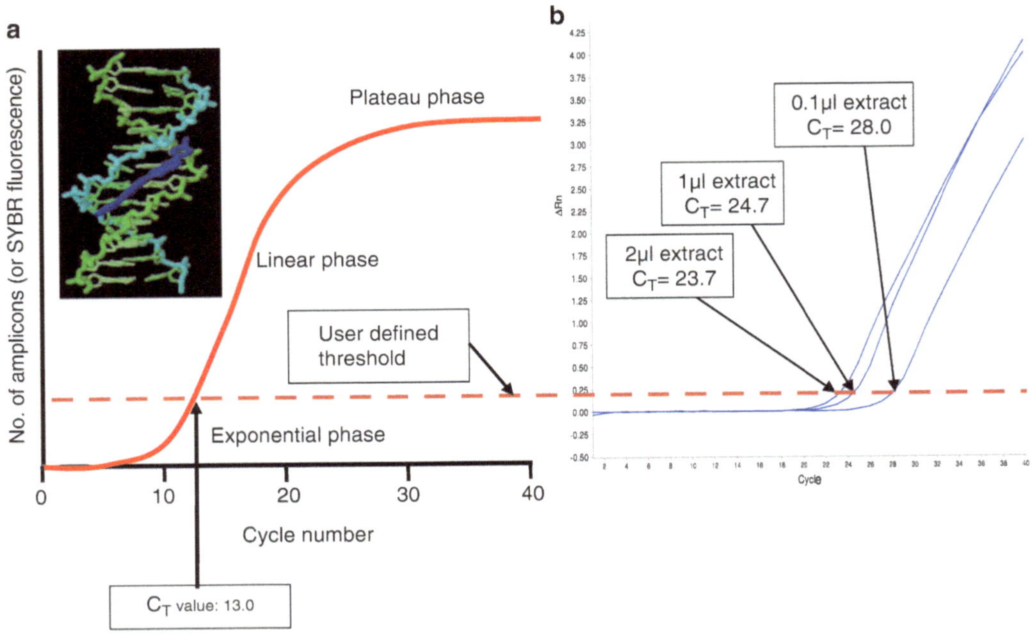

Fig. 1. (**a**) A sigmoidal PCR amplification curve, depicting the exponential, linear, and plateau phases of the reaction, together with a schematic representation of the extrapolation of C_T values. The insert depicts a SYBR Green molecule (*dark blue*) bound to the minor groove of a DNA duplex. (**b**) A qPCR assay in which 2, 1, and 0.1 µL (via dilution) have been placed in the reaction. The C_T values shift in accordance with the dilution factor.

published literature on qPCR involves research on the quantitation of gene expression via RNA, which has limited applicability to low copy number aDNA templates. This chapter will focus exclusively on the use of SYBR Green-based qPCR as opposed to the probe-based TaqMan systems, as we have found this method to be the more sensitive and cost-effective assay.

qPCR focuses on imaging the amount of amplicon present during the exponential phase of the PCR using a fluorescent dye (SYBR Green) that binds specifically in the minor groove of double-stranded DNA (Fig. 1a). All qPCR systems are equipped with a camera that captures the fluorescent output of the reaction at every cycle. In the exponential phase (Fig. 1a), the amplicon concentration, as measured by SYBR Green fluorescence, correlates with the original level of input DNA. The output from qPCR assays is expressed as a cycle threshold value, C_T (also called C_q), which represents the number of cycles taken to reach a certain user-defined threshold value in signal strength (Fig. 1a). If the threshold is fixed, the C_T values estimated from different PCRs can be interpreted as a relative measure of template copy numbers between DNA extracts. The threshold value needs to be defined at a level where the amount of fluorescence is both detectable by the camera and is above the background level of SYBR Green binding in the reaction (Fig. 1b).

Since the C_T value images the PCR in the exponential phase, it should reflect the level of input DNA (in the absence of inhibitors and assuming 100% PCR efficiency—see below). As depicted in Fig. 1b, if 2 μL of an aDNA extract is aliquoted into a reaction and gives a C_T of 23.7 cycles, then adding only 1 μL should result in a C_T of 24.7 cycles. Likewise, adding only 0.1 μL of the same DNA extract (via dilution) should yield a C_T of 28.0 cycles (a shift of 3.3 cycles is expected from a 1/10 dilution: $2^{3.3} = 10$).

The applications of qPCR in aDNA research are many, but fall into five main categories:

1. *Optimization of aDNA recovery*: Using qPCR C_T values for absolute quantitation using either a standard or relative CT value (see studies on bone (4) and eggshell (5) respectively), it is possible to explore different extraction protocols to maximize the recovery of DNA from ancient substrates.
2. *Detection of PCR inhibitors*: The copurification of compounds that adversely affect PCR is a common problem in aDNA research, especially in sediments (e.g., humics and tannins). By performing a serial dilution of aDNA extracts, it is possible to detect inhibition as the absence of expected C_T shifts as outlined above. In many instances, the apparent number of amplifiable template molecules may actually increase (resulting in a lower C_T value) as the aDNA extract is diluted, as this also reduces the concentration of inhibitors. Extracts can also be spiked with an unrelated primer set and synthetic template to detect inhibition (internal PCR controls, or IPCs). If the IPC is positive and amplifies at the same C_T with and without the spiked aDNA extract, PCR inhibition is not adversely affecting the reaction.
3. *Rapid assessment of preservation*: qPCR data can enable ranking samples in order from the best preserved to the poorest preserved by comparing C_T values. This has benefits in prioritizing samples, identifying environments with favorable DNA preservation, and examining the effects of postmortem DNA diagenesis in samples or stored extracts (6).
4. *aDNA authentication*: qPCR data can serve as a useful (but by no means definitive) means to assess data fidelity. If a reaction has amplified from a very small number of starting template molecules, that reaction is likely to be more susceptible to contamination. Moreover, when dealing with very poor DNA preservation and low copy numbers of target DNA fragments, a larger fraction of the available templates may display miscoding lesions as the result of postmortem DNA damage (7).
5. *aDNA library quantification*: aDNA genomic libraries constructed for high-throughput sequencing (HTS) applications (e.g., Roche 454, ABI SOLiD, and Illumina Solexa) need to be accurately quantified. Unlike modern DNA, the quantum

of template is a limiting factor. qPCR can be used to quantify libraries both pre- and post-enrichment (8).

In this chapter, we discuss the construction of SYBR Green qPCR assays and standards. We provide details of how to assemble the qPCR components rather than using commercially available qPCR master mixes, which offer less flexibility in modifying reaction chemistry. In our laboratory we use qPCR to obtain information about levels of inhibition in the aDNA extract and relative preservation between samples (5, 9, 10). By designing qPCR primer pairs in phylogenetically informative regions, amplicons generated in the qPCR assay can be sequenced directly.

2. Materials

1. Laboratory equipment: A standard set of equipment is required including calibrated pipettes, aerosol resistant tips, gloves, lint-free wipes, aluminum foil, a vortex mixer, and a microcentrifuge suitable for 1.5-mL and PCR strip tubes.
2. Plasticware: 1.5- and 0.5-mL microcentrifuge tubes are needed for reagent preparation. Thin-walled PCR strip tubes, with optical lids suitable for qPCR are required (see Note 1).
3. PCR primers suitable for qPCR: 10 μM aliquots of both the forward and reverse PCR primers suitable for use in a qPCR assay (see Note 2).
4. qPCR standard: 100 μM solution of a synthetic oligonucleotide (see Notes 3 and 4; Fig. 2a).
5. A solution of 10 mM Tris (pH 8.0) to use when diluting DNA extracts or standards.
6. SYBR Green dye: A 1:2,000 dilution of SYBR Green I (10,000× Nucleic Acid Gel Stain; Invitrogen catalogue number S7563) in molecular biology grade DMSO (see Note 5).
7. PCR reagents:

   GeneAmp 10× PCR Buffer (Applied Biosystems).

   dNTPs 25 mM of each nucleotide.

   25 mM $MgCl_2$ solution.

   AmpliTaq Gold DNA polymerase (Applied Biosystems—see Note 6).

   Molecular biology grade H_2O.

   Molecular biology grade bovine serum albumin (BSA) 10 mg/mL.

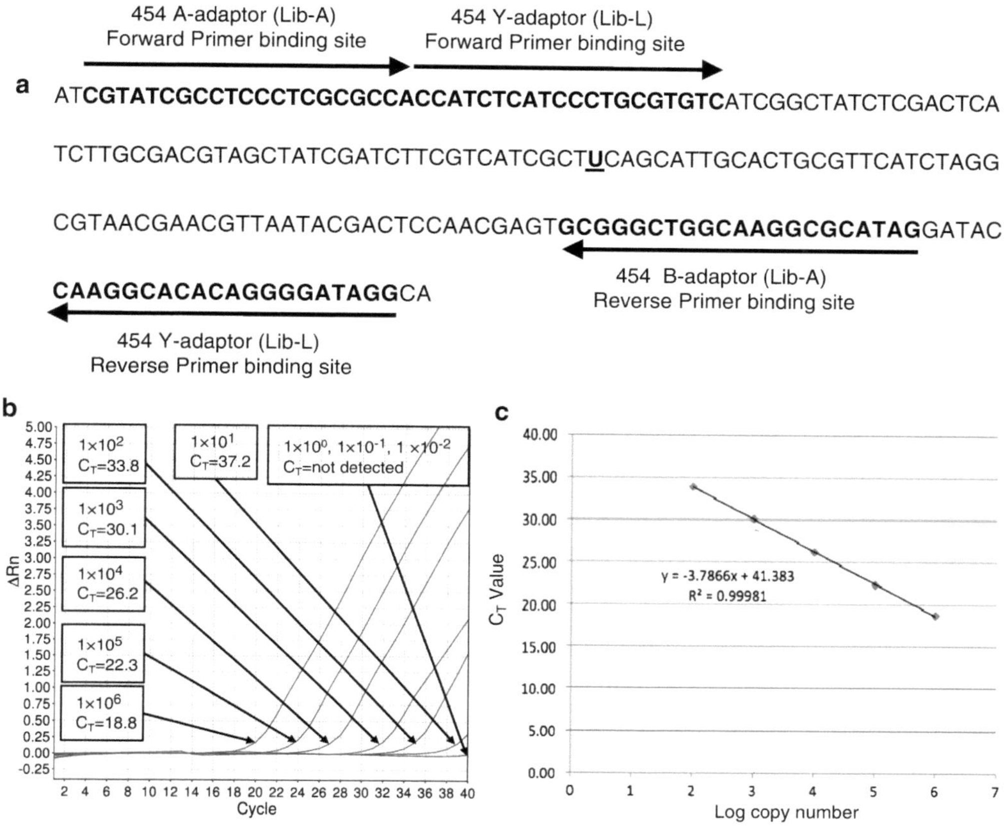

Fig. 2. (**a**) Synthetic oligonucleotide (5′ to 3′) ordered for absolute quantitation of aDNA Roche 454 libraries. This is a universal standard as it has primer binding sites for both 454 A/B primers (Lib-A, marked by arrow) as well as Y-adaptor primers (Lib-L, marked by arrow). A single Uracil residue (U, underlined) is included in the middle of the sequence which allows the oligo to be inactivated by UNG in the event of contamination. (**b**) A qPCR assay of the synthetic oligo standard pictured in (**a**), using the A/B (Lib-A) primers. A 1/10 dilution series from 10^6 to 10^{-2} copies was placed in the reaction tubes and the C_T values recorded. (**c**) The C_T values from (**b**) are plotted against the log-transformed copy number to generate a regression line that can be used to extrapolate samples with unknown concentration from any C_T value. The optimal emPCR bead-to-template ratio based on your standard will need to be determined empirically when using this standard.

3. Methods

3.1. Preparation of qPCR Standards

Due to the high concentrations of standard templates involved, make up the synthetic oligonucleotide standard in a location that is separate from all molecular biology activities, and also separate from the aDNA clean room.

1. Vortex the 100 μM stock oligonucleotide standard. Add 10 μL of this stock to 590 μL of 10 mM Tris pH 8.0. This tube will (according to Avogadro's number) contain 1×10^{12} copies of single-stranded template per microliter.
2. In a series of 1/10 serial dilutions in 10 mM Tris pH 8.0 (being mindful of pipetting errors and contamination), obtain a set of standards that contains 1×10^6–1×10^{-2} copies per microliter.

3. Test the synthetic standard using the qPCR assay (Fig. 2b). Add 2 μL of the template (of the 1×10^{6}–1×10^{-2} copies per μl stock solutions) to each reaction to obtain the concentrations that correspond to double-stranded DNA (see Note 7).
4. The standard curves should respond in a clear quantitative manner (approximately 3.3 cycle increase per 10× dilution) and should fail to amplify below zero copies (Fig. 2b, see Note 8). Perform the standard dilution series a number of times independently to ensure reproducibility.
5. Plot the $\log_{10}$ transformed concentration values against C_T, this should generate a linear regression line with a high R^2 value (Fig. 2c). This can be used to extrapolate the DNA copy number from the C_T values of any unknown samples. A calculation of primer efficiency should also be carried out—see http://www.gene-quantification.info/ and references therein (see Note 9).

3.2. Setting Up the qPCR Reaction

1. Set up the qPCR assay in a DNA clean room (see Note 10). Due to contamination risks, oligonucleotide standards should always be added in another laboratory after the reaction tubes containing aDNA extracts have been sealed in the clean room.
2. The following recipe provides a generic template for a 25 μL qPCR mix, but the reaction chemistry may need to be optimized (especially Mg^{2+} and primer concentrations) depending on the specificity of the assay. Mix the following reagents together (the "master mix") in a 1.5-mL tube. The recipe below is shown for one PCR reaction and will need to be scaled up to the number of required reactions.

GeneAmp 10× PCR buffer	2.5 μL
dNTP's, 25 mM of each nucleotide	0.25 μL
25 mM $MgCl_2$ solution	2.5 μL
AmpliTaq Gold DNA polymerase	0.25 μL (or 1–2 units)
Forward PCR primer (10 μM)	1 μL
Reverse PCR primer (10 μM)	1 μL
SYBR Green dye	0.6 μL
BSA	1 μL
Molecular biology grade H_2O	13.9 μL
Template DNA (*omit from master mix!*)	2 μL (to be added later)
Final volume	25 μL

3. Ensure that all components are well mixed, especially the *Taq* polymerase, which is stored in high concentrations of glycerol and can settle at the base of the tube.

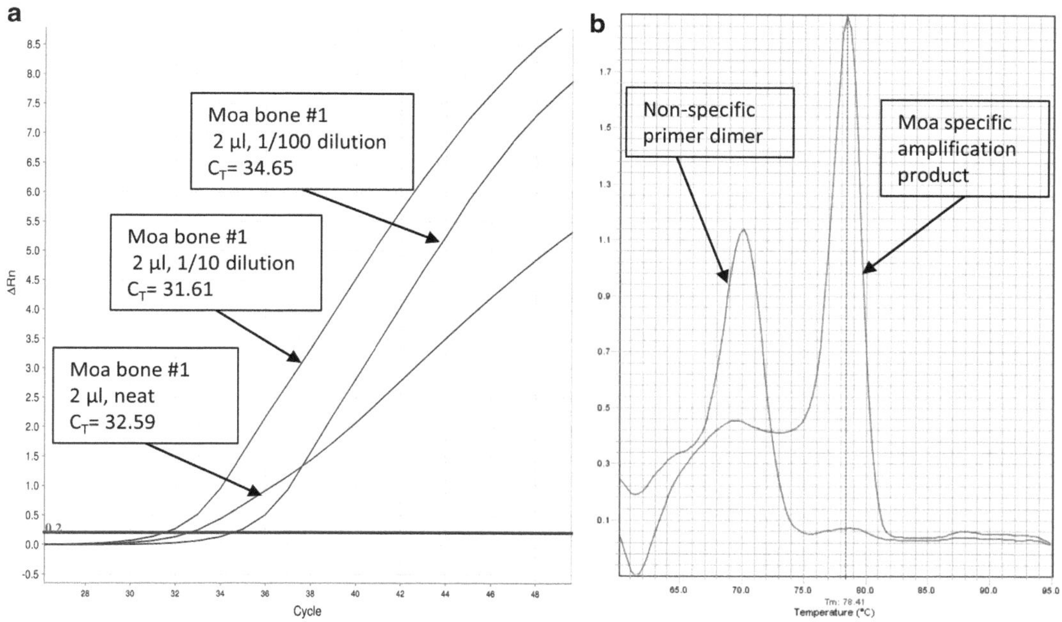

Fig. 3. (**a**) qPCR assay conducted on a moa bone that demonstrates the effect of PCR inhibitors. The "neat" extract is clearly compromised in this assay. In many instances, inhibited reactions fail to demonstrate any amplification. (**b**) A melt-curve analysis of the assay shown in (**a**): the curve enables the user to easily discriminate between reactions yielding *bona fide* products and reactions that have been compromised by dimer formation. In many instances, the melt analysis will circumvent the need to run the PCR products on an agarose gel.

4. Carefully aliquot 23 µL of master mix into each reaction tube. Seal all the reaction tubes once 23 µL have been dispensed.
5. Once sealed, identify the reaction tubes in an 8-strip by pen marks on the hinge region. It is *not* advisable to write on the top of the optical lid as the dye in the marker pen may interfere with the fluorescence detection by the camera.
6. If not freshly extracted (see Note 11), thaw the aDNA extracts and let them equilibrate to room temperature. It may be necessary to construct dilution series (in 10 mM Tris, pH 8.0) on the DNA extracts to test the quantitative response, for example to make sure that inhibition is not affecting the results (e.g., Fig. 3a).
7. In a separate area of the clean room (see Note 10), add 2 µL of template DNA to each reaction tube. Open tubes one at a time and close immediately after the addition to minimize cross-contamination.
8. Transport the tubes wrapped in foil to PCR facility.
9. If required, add the qPCR standards or a positive control to the appropriate tubes. If adding a standard, it is advisable to begin by adding the most dilute DNA concentration (to minimize the

risk of contamination from the highest concentration). Ensure the lids of all tubes are well sealed (see Note 12).

10. Pulse-spin the reaction tubes.

3.3. Thermocycling on a qPCR Platform

There are many qPCR platforms available, all of which are equipped to detect SYBR Green that has excitation and emission maxima of 494 and 521 nm, respectively. Follow the manufacturer's instructions regarding loading and programming the specific qPCR platform. During the loading process avoid the use of powdered gloves as the powder can adversely affect the fluorescent detection.

1. Place reaction tubes into the thermocycler block, making sure that no air bubbles are present or that the reaction contents have not splashed up the side of the tube.
2. Once loaded, wipe the optical lids of the tubes with a lint-free tissue such as a kim-wipe.
3. Within the qPCR platform software, specify:

   Which wells to monitor.

   Which wells contain no template controls (NTCs).

   If relevant, which wells contain standards.

   Detection of SYBR Green dye (and passive reference dyes if relevant).
4. Program the thermocycling conditions for either a two-step PCR protocol (e.g., 95°C/60°C) or three-step PCR protocol (e.g., 95°C/60°C/72°C) (see Note 13) for 40 cycles. Importantly, the annealing temperature must be optimized specifically for each primer set to maximize efficiency and minimize dimer.
5. Instruct the machine to execute a melt curve analysis (see Note 14) at the end of the 40 cycles. An example of a melt curve can be found in Fig. 3b.
6. If amplicons from the qPCR are to be used in any downstream applications (i.e., cloning or HTS amplicon library builds), a final 72°C step (5–10 min) can be included to promote terminal adenylation.
7. Analyze the output of the qPCR run (see Note 15). Most qPCR software packages will automatically determine C_T values and the threshold level that generates these values. The threshold value needs to be standardized if C_T values are to be compared across different runs. By comparing C_T values from a dilution series, it is possible to detect inhibition (Fig. 3a) and, in the absence of inhibition, to determine which extracts have higher yields of DNA. Likewise, by using a standard curve (e.g., a standard for Roche 454 library quantification) it is possible to determine the absolute number of molecules (Fig. 2a, b) for an aDNA library.

8. The publication of qPCR data (principally for gene expression studies) has recently undergone its own set of authentication criteria referred to as MIQE, or minimum standard for the provision of information for qPCR experiments (11). Users of qPCR in aDNA research should be aware of the MIQE guidelines as they may, in time, impact on the ability to publish qPCR data.

4. Notes

1. The choice of appropriate plasticware is very important in qPCR. The use of 96-well plates and 8-strips with independent 8-strip lids may be problematic in an aDNA context as reactions cannot be sealed individually. We strongly advise the use of 8-strips with the attached lids (such as SnapStrip II PCR tubes, Anachem) so that each reaction tube can be opened and closed one at a time.
2. A SYBR Green-based qPCR assay relies on a highly efficient primer set that ideally generates amplicons of less than 250 bp (an arbitrary cut-off that is meant to ensure that all assays function efficiently). Careful consideration must be given to ensure that the PCR reaction results in a minimum of homo- and heterodimers (as the dimers will also bind SYBR Green). Researchers can refer to a number of online tools to assist in primer design. It is well accepted that different primer sets have different sensitivities and hence different quantitation limits (the point at which the shifts in C_T values cease to respond as expected). It is advisable that a number of primer combinations are used and tested across a range of dilutions to find the assay that is best suited to the application.
3. In some instances, absolute quantitation (exact copy numbers) may be required, and there are a number of ways to generate a standard curve. Standards can be useful in determining the qPCR assay sensitivity, defined as the absolute number of template molecules that a given qPCR assay can reproducibly detect. While amplicons or plasmids can be used as standards (using A260 or PicoGreen, Invitrogen, staining to determine concentrations), in our experience synthetic oligonucleotides are more reliable and reproducible. However, in other aDNA applications, absolute quantitation may not be required. For instance, the relative amount of DNA in each extract can easily be compared using only C_T values (free of inhibition) without an estimated absolute number. In fact, the considerable contamination risk involved with the use of high copy number standards might cancel out some of the benefits.

4. A synthetic oligonucleotide that is approximately the same amplicon length as your assay should be employed as a synthetic standard. Many providers can synthesize long oligonucleotides, and HPLC purification is not normally required. The primer sequences should be incorporated into the 5 and 3′ ends (see Fig. 2a). The internal sequence can mimic the target sequence but should include an identifiable run of nucleotides (e.g., an insertion) so that it can be differentiated from that of your assay target. For contamination control purposes, it is also advisable to include uracil residue(s) in the middle of the sequence. This enables the standard to be inactivated by UNG if required (see Fig. 2a).
5. Make aliquots of the SYBR reagent in a clean room to avoid freeze–thaw cycles. Each PCR will require ~0.6 μL of the reagent (see below). The SYBR is stable for approximately 1 year, and should be stored in the dark wrapped in foil. The 10,000× Gel stain is a cost-effective means by which to obtain SYBR Green. Note that the final concentration of SYBR Green in the PCR is not 1× of the original 10,000× stock. The described concentration that equates to an approximate concentration of 0.12× is optimal for use in qPCR. A Rox dye (Invitrogen catalogue number 12223–012) can also be used as an internal qPCR normalizer if required. Dilute the Rox dye 1:500 in DMSO, and use approximately 0.3 μL per 25 μL reaction (this will need to be optimized for your qPCR platform).
6. We prefer AmpliTaq Gold DNA polymerase (Applied Biosystems) for our aDNA qPCR assays, as the enzyme performs well on aDNA templates (4). However, other commercially available hot-start *Taq* polymerases may be equally efficient. We have trialed a number of proofreading polymerases both alone and as part of blends, all of which have performed poorly in qPCR assays.
7. Because each oligonucleotide standard is single-stranded, you will need to add twice the determined volume to obtain a double-stranded equivalent. For example, 2 μL of a 1×10^6 standard will be required to simulate 1 μL of a 1×10^6 double-stranded standard.
8. The sensitivity and detection limits will differ between primer sets. However, no amplification should be observed at C_T values representing less than one standard template molecule. If this happens, the absolute copy number derived from the standard curve is not reliable. In our experience with oligonucleotide standards, the fewest number of copies that can be detected is between 2 and 20, and a reproducible quantitative result can be achieved at around 50–100 copies.

9. All primer sets will have different sensitivities, detection, and quantitation limits, and it is important to explore these variables to ensure the fidelity and validity of your qPCR assay.
10. Within the aDNA clean room, it is advisable to separate the area (and pipettes) where PCR setup occurs from the area where aDNA extract is added to the PCR tubes (e.g., using two different PCR hoods or glove boxes). This setup will minimize the risk that reagents and equipment become cross-contaminated with DNA.
11. The preferred time to conduct qPCR is when aDNA is freshly extracted. This maintains consistency between extracts, as storage time and freeze–thaw cycles will influence the number of amplifiable templates. Long-term storage of the DNA extract can result in DNA degradation and/or binding of DNA to the surfaces of plastic tubes.
12. Ensuring a good seal may seem trivial, but one problematic outcome of having a non-sealed tube (or plate seal) is that the SYBR dye will volatilize and condense on the camera (or in the well). This will generate anomalous qPCR results, and may also damage the qPCR detection unit. For this reason it is not advisable to repeatedly use the same wells in a thermocycler block, especially during troubleshooting.
13. The decision to implement a two-step or three-step qPCR is assay-specific. Many manufacturers advocate a two-step strategy, citing evidence that with a two-step procedure more reliable data is obtained during the exponential phase. However, some qPCR assays may simply not be suited to this approach.
14. Briefly, a melt curve analysis is an endpoint assay where reactions are heated to 60°C, imaged for SYBR, and then increased to the next temperature (typically 1°C steps) until 95°C. When the double-stranded DNA in the reaction denatures, a drop in fluorescence is detected. Therefore, melt-curves can be a very valuable tool in the detection of primer dimer and nonspecific amplification products, as these will tend to denature at different temperatures (i.e., if they have different lengths and/or a different GC-content) than the target sequence (Fig. 3b). Melt curves can eliminate the need to run an agarose gel, thereby saving time and decreasing the concentration of amplicon aerosols circulating in the laboratory.
15. An in-depth discussion of the analysis of qPCR outputs is beyond the scope of this chapter. The C_T (or absolute numbers if using a standard) generated by qPCR is only reliable if the assay is validated (11). The presence of inhibition, primer dimer, nonspecific amplification, and poor laboratory techniques will compromise the data fidelity. Once a DNA isolation protocol has been optimized, the C_T values can be used to

identify (or discard) samples that have, for example, a relatively high, medium, or low template copy number. It is often the case (at least with bone and eggshell) that samples showing high copy number also yield longer amplicons, work for nuclear DNA targets, and show less allelic drop-out when amplifying nuclear microsatellites or sex-linked loci (9, 10).

Acknowledgments

MB was supported by the Australian Research Council as a Future Fellow (FT0991741). We thank Jayne Houston and James Haile for helpful discussions and Beth Shapiro for valuable editorial inputs.

References

1. Bustin SA (2004) A-Z of quantitative PCR. International University Line, La Jolla, CA
2. Bustin SA (2000) Absolute quantification of mRNA using real-time reverse transcription polymerase chain reaction assays. J Mol Endocrinol 25:169–193
3. Bustin SA, Benes V, Nolan T, Pfaffl MW (2005) Quantitative real-time RT-PCR—a perspective. J Mol Endocrinol 34:597–601
4. Rohland N, Hofreiter M (2007) Comparison and optimization of ancient DNA extraction. Biotechniques 42:343–352
5. Oskam CL, Haile J, McLay E, Rigby P, Allentoft ME, Olsen ME, Bengtsson C, Miller GH, Schwenninger JL, Jacomb C, Walter R, Baynes A, Dortch J, Parker-Pearson M, Gilbert MT, Holdaway RN, Willerslev E, Bunce M (2010) Fossil avian eggshell preserves ancient DNA. Proc Biol Sci 277: 1991–2000
6. Pruvost M, Schwarz R, Correia VB, Champlot S, Braguier S, Morel N, Fernandez-Jalvo Y, Grange T, Geigl EM (2007) Freshly excavated fossil bones are best for amplification of ancient DNA. Proc Natl Acad Sci USA 104:739–744
7. Gilbert MT, Binladen J, Miller W, Wiuf C, Willerslev E, Poinar H, Carlson JE, Leebens-Mack JH, Schuster SC (2007) Recharacterization of ancient DNA miscoding lesions: insights in the era of sequencing-by-synthesis. Nucleic Acids Res 35:1–10
8. Meyer M, Briggs AW, Maricic T, Hober B, Hoffner BH, Krause J, Weihmann A, Paabo S, Hofreiter M (2008) From micrograms to picograms: quantitative PCR reduces the material demands of high-throughput sequencing. Nucleic Acids Res 36(1):e5
9. Allentoft M, Schuster S, Holdaway R, Hale M, McLay E, Oskam C, Gilbert MT, Spencer P, Willerslev E, Bunce M (2009) Identification of microsatellites from an extinct moa species using high-throughput (454) sequence data. Biotechniques 46:195–200
10. Allentoft ME, Bunce M, Scofield RP, Hale ML, Holdaway RN (2010) Highly skewed sex ratios and biased fossil deposition of moa: ancient DNA provides new insight on New Zealand's extinct megafauna. Quat Sci Rev 29:753–762
11. Bustin SA (2010) Why the need for qPCR publication guidelines?—The case for MIQE. Methods 50:217–226

Chapter 17

Multiplex PCR Amplification of Ancient DNA

Mathias Stiller and Tara L. Fulton

Abstract

Multiplex PCR allows the simultaneous amplification of up to dozens of target fragments in a single PCR. It is therefore a powerful tool to obtain many kilobases of continuous sequence from minute amounts of ancient DNA (aDNA), which usually must be amplified in multiple short and overlapping fragments. Because significantly less template is required compared to amplifying each fragment separately, multiplex PCR is particularly beneficial when the fossil material itself, or access to the fossil material, is limited. The recently refined two-step multiplex PCR protocol consists of a first-step reaction (the actual multiplex PCR) that then acts as the template for the second-step PCR. During the second step, nested primers are used in individual amplification reactions. Although the same set of primers can be used in both steps, using a nested set in the second step adds an additional level of selectivity and specificity, minimizing PCR artifacts. This is particularly important when complex mixtures of template DNA, such as aDNA extracts, are amplified.

Key words: Polymerase chain reaction, Two-step multiplex PCR, Singleplex PCR, Monoplex PCR, Overlapping fragments, Nested primer, Ancient DNA

1. Introduction

The polymerase chain reaction (PCR) offers specificity and single-molecule sensitivity, making it an excellent method for analyzing ancient DNA (aDNA). However, each PCR requires at least one template molecule of the desired genomic region. Depending on DNA preservation, more or less DNA extract will be required to begin each amplification reaction. Multiple amplifications of short, overlapping fragments may be necessary to reconstruct long, informative DNA sequences. This process is performed in replicate to account for possible sequence errors due to miscoding lesions in the template molecule (1). Ignoring economical constraints, the total number of PCR amplifications that can possibly be performed will depend on the amount of the sample that exists and the

Beth Shapiro and Michael Hofreiter (eds.), *Ancient DNA: Methods and Protocols*, Methods in Molecular Biology, vol. 840, DOI 10.1007/978-1-61779-516-9_17, © Springer Science+Business Media, LLC 2012

preservation of that sample. When analyzing unique or rare samples (such as hominids or carnivores), the amount of sequence data that can be obtained from a given specimen may be limited.

Multiplex PCR was first established in 1988 to simultaneously amplify multiple loci in the human dystrophin gene (2) and has been used subsequently in a variety of applications including pathogen identification, gender screening, linkage analysis, template quantitation, genetic disease diagnosis, forensics, and population genetic studies (see (3) for a comprehensive overview). Although low-number multiplex PCR has been used previously to amplify aDNA (4, 5), the recently developed two-step approach (6, 7) is the only approach to date that is able to overcome some of the major obstacles of aDNA research.

Using the two-step approach, Krause and colleagues (6) were able to reconstruct the complete mitochondrial genome (16,770 base pairs, bp) of a 12,000-year-old wooly mammoth using the equivalent of only 200 mg of bone, based on two primary (first-step) multiplex reactions of 23 primer sets each. Since 2006, the approach has been used successfully to amplify both mitochondrial (8–13) and nuclear aDNA (14, 15) from various extinct species. Recently, the method was coupled with a new method of generating barcoded sequence libraries for sequencing on next generation high-throughput sequencing platforms (i.e., Roche's 454 or Illumina's Solexa; see Chapter 20 (16)).

The first step of the two-step amplification strategy is the multiplex PCR. In this step, every second fragment of a series of overlapping fragments is amplified by adding the appropriate primer pairs into one of two separate, nonoverlapping mixtures, commonly referred as to the "odd" and the "even" set. Importantly, overlapping fragments must not be amplified in the same PCR, because the forward primer of fragment 2 can pair up with the reverse primer of the upstream fragment 1. The odd set therefore includes the primer pairs for fragments 1, 3, 5, and so on, and the even set includes the primers amplifying fragments 2, 4, 6, and so on. Thus, only two PCRs need to be performed, compared to a single reaction for each primer pair in standard PCR.

A dilution of this first-step multiplex amplification reaction serves then as template for the subsequent second-step amplifications. In the second step, each of these individual PCRs (also called simplex, singleplex or monoplex PCRs) includes only a single primer pair to be amplified. Because the starting template is the amplified product from the first-step reaction, rather than the original DNA extract, the amount of the original DNA extract that was necessary to amplify all of the targeted region (or regions) is dramatically reduced.

2. Materials

2.1. PCR Reagents

All reagents and plastic consumables must be sterile (DNA and DNAse free) and of molecular biology grade or similar.

1. Deoxynucleoside triphosphates (dNTPs) of 100 mM each, combined in equal volume to yield a dNTP mix of 25 mM each dATP, dGTP, dTTP, and dCTP.
2. DNA polymerase + buffer supplied with polymerase, usually AmpliTaq Gold (see Note 1).
3. Magnesium ions supplied separately with polymerase usually as $MgCl_2$ or $MgSO_4$.
4. Forward and reverse primers in 100 μM stock solutions (see Note 2) as well as diluted to 10 μM.
5. Serum albumin (SA), e.g., bovine (BSA) or rabbit serum albumin (RSA), prepared as 10 mg/mL solution in sterile water (see Note 3).
6. Barrier/filter tips and PCR reaction tubes/plates.
7. Thermocycler with heated lid.
8. DNA template.
9. Flowhood or PCR-free region in the modern laboratory for the second-step PCR setup.

2.2. Agarose Gel Visualization and PCR Purification

1. 2% Agarose gel.
2. 1× TAE (Tris–acetate–EDTA) or 0.5× TBE (Tris–borate–EDTA) running buffer, available commercially.
3. 6× loading dye (0.25% Orange-G (TCI), 0.1875% xylene cyanol (IBI Scientific), 30% glycerol).
4. DNA ladder: The ladder can be diluted with TE buffer: 125 μl (0.25 μg) prepared ladder + 1125 μL TE + 250 μL 6× loading dye (supplied with ladder).
5. Agarose gel electrophoresis rig and power supply.
6. Ethidium bromide (EtBr) and a UV transilluminator.
7. Your preferred commercial PCR purification kit: Qiagen, Millipore, ExoSAP, Agencourt AMPure XP.

3. Methods

3.1. Primer Design

1. Design a series of overlapping primer sets that do not vary by more than 3°C in their melting temperatures, as they will all be cycled at the same temperature. The targeted regions should

generally range in length between 100 and 300 bp excluding primers. Divide the primer sets into the "odd" and "even" sets. These sets can be used for both the first and second-step PCRs (see Note 4).

2. To maximize specificity and selectivity for the desired target fragments, a second set of primers can be designed for use in the second-step PCRs (instead of using the same set of primers in both steps). In this case, one or both primers used in the second-step amplifications would be partially or even fully nested within the target fragment (see Note 5).

3.2. First-Step PCR Setup

1. In an aDNA clean room, prepare the primer sets by mixing all the forward and reverse primer pairs of the "odd" set together to obtain a working primer solution in which each primer will have a final concentration of 1 μM in the primer solution (see Note 2). Do the same for the "even" set.
2. Prepare the aDNA multiplex PCR master mix (add all reagents except the DNA template) as outlined in Table 1. Prepare sufficient master mix for all samples plus a PCR negative control reaction for every 8–10 sample reactions (see Note 6).
3. Dispense the master mix into the PCR strip tubes or plates, and then close all the lids.
4. Add DNA template individually to each reaction tube, opening only one tube (or row of tubes, if using multichannel pipettes) at a time to avoid cross-contamination.
5. Before leaving the clean room, set up the second-step PCR reaction mix as described in Subheading 3.3. The first-step

Table 1
First-step multiplex PCR setup

Reagent	Volume (μL) per sample	Final concentration in reaction
Water (add to make 20 μL final reaction volume)		
10× buffer	2	1×
25 mM $MgCl_2$	2–3.2	2.5–4 mM
BSA or RSA (10 mg/mL)	2	1 mg/mL
dNTPs (25 mM each)	0.2	0.25 mM each
Primer mix (1 μM each primer)	3.0	0.15 μM each primer
AmpliTaq Gold (5 U/μL)	0.4	2 U
Template	1–5	

reactions can be kept in cold while you prepare these reactions (see Note 7).

6. After the first-step PCR is completely set up and the second-step PCR is missing only the template DNA (this will be the result of the first-step PCR), proceed to the modern lab. Cycle the first-step reactions using an initial step of 95°C for 12 min to activate the hot-start polymerase, followed by 30–40 cycles of denaturation for 30 s at 94°C, primer annealing for 30 s at the required annealing temperature determined by your primer set, and elongation for 40 s at 72°C. Finish the cycling protocol with a final extension step at 72°C for 4 min (see Note 8).
7. After the cycling is finished, dilute a part of the first-step reactions 1:20–1:50 (depending on how many second-step reactions you are planning to set up) with clean water (see Notes 9 and 10). This will be the template for the second-step PCR.

3.3. Second-Step PCR Setup

1. Plan the plate layout. If sufficient samples are being processed that multichannel pipettes will be used, set up the second-step PCR so that the first-step reactions can be diluted and dispensed straightforwardly using the multichannel pipette (see Note 11).
2. In the clean room, set up the second-step reaction master mixes (also called monoplex, simplex, or singleplex PCRs) for each single primer pair separately following the recipe outlined in Table 2. Include sufficient mix for at least one PCR negative control per 8–10 first-step reactions to monitor any contamination introduced during the process of diluting the first-step

Table 2
Second-step multiplex PCR setup

Reagent	Volume (μL) per sample	Final concentration in reaction
Water (add to 20 μL)		
10× buffer	2	1×
25 mM $MgCl_2$	2–3.2	2.5–4 mM
BSA or RSA (10 mg/mL)	2	1 mg/mL
dNTPs (25 mM each)	0.2	0.25 mM each
Forward primer (10 μM)	1.5	0.75 μM
Reverse primer (10 μM)	1.5	0.75 μM
AmpliTaq Gold (5 U/μL)	0.1	0.5 U
Template (dilution of first-step PCR)	5	

PCR or when adding the dilutions as a template to the second-step reactions. No template is added yet.

3. Dispense the master mix and seal the tubes or plates. Take this master mix to the modern lab (see Note 12).
4. After the first-step PCR is complete and the reactions are diluted (Subheading 3.2, step 7), add the template to the second-step reactions in a laminar flow hood or area free of post-PCR processing (like agarose gel visualization) that is *not* the aDNA lab (see Note 13).
5. Perform the second-step PCR reactions using the same cycling conditions as used for the first-step PCR in Subheading 3.2, step 6 (see Note 8).

3.4. Agarose Gel Visualization and PCR Purification

1. Prepare a 2% or higher concentration agarose gel in TAE (or TBE). Ethidium bromide (EtBr) may either be included in the buffer, the gel, or applied as a post-stain. EtBr is a mutagen and must be handled with care (see Note 14).
2. Run out 3 μL of the completed second-step PCR and the PCR negative controls from the first-step reactions (see Note 15). Visualize with UV light. If bands of the expected length are present in the first-step negative(s), repeat the first-step PCR for the particular primer set. If bands are present in the second-step negative(s), repeat the second-step PCR for the particular primer pair.
3. If the second-step PCR yields a single clean band, purify the remaining 17 μL of the reaction using your favorite commercial PCR purification system (see Note 16).
4. The PCR products can now either be cloned or directly sequenced using traditional Sanger sequencing or processed with high-throughput sequencing methods, potentially incorporating a barcoding step.

4. Notes

1. It is easier to optimize reactions when the magnesium is not included in the buffer, as changing the concentration of magnesium is a useful optimization protocol.
2. If you intend to use more than 100 single primers (i.e., 50 primer pairs) in a single primer set, a more concentrated stock solution is required, since the final concentration of each primer within the primer set should be 1 μM.
3. Although serum albumin (SA) is not required for PCR, it is almost always used in aDNA PCR.

4. Due to the abundance of short aDNA template molecules in the extracts, the overlaps between adjacent fragments may be preferentially amplified compared to the desired long fragments and thus outcompete them in the course of the PCR reaction. In case of a very tight tiling of very short target fragments, three or four different primer sets, rather than the "odd" and "even" sets described here, may be necessary.
5. This strategy is particularly useful when attempting to amplify nuclear loci. When targeting mitochondrial fragments, the identical primers from the first-step amplifications are generally used in the second-step PCRs.
6. As in standard PCR, it may be beneficial to add carrier DNA (e.g., lambda phage DNA or DNA from another nontarget species) in the PCR negative controls to exclude potential carrier effects that may prevent detection of contamination. PCR positive controls should be avoided if possible.
7. If the mix for the second-step PCR is to be prepared on a different day, the first-step PCR can be stored in the freezer in the modern lab.
8. The cycling conditions provided are specific to AmpliTaq Gold (Applied Biosystems). If a different polymerase is used, adjust the cycling conditions according to the manufacturer's recommended activation and extension temperatures.
9. The water used to dilute the first-step reactions should be sterile. We recommend bringing water in strip tubes or in plates directly from the clean room, where it can be prepared while setting up the PCR. This avoids carry-over contamination in the second-step reactions due to the use of potentially contaminated water from the post-PCR facilities.
10. Performing two second-step reactions from the same first-step template is not a replication of the PCR, since both second steps start from the same initial template molecule(s) amplified during the first-step reaction. Two independent, first-step amplification reactions must be performed to meet the criteria of independent replication.
11. For example, if eight samples are processed in the first PCR, set up the second PCR so that sample 1 is the template for 12 singleplex reactions in row A of a 96-well plate, sample 2 is in row B, etc. Thus, singleplex primer set 1 will be used for column 1 on the plate, primer set 2 in column 2, etc.
12. Using a hot-start polymerase like AmpliTaq Gold (Applied Biosystems) makes it possible to set up the second-step reactions in the clean room immediately after the first-step reactions are prepared. The second-step reaction mix can then be stored with sealed lids outside the clean room until the first-step reactions are finished, and amplified products from the

first step can be added to the prepared second-step reactions at that point.

13. As the template for the second-step PCR is already a PCR-amplified product, it is critical that it is not brought back into the clean room, which must stay free of PCR product. However, the reactions are still highly sensitive to contamination and must be prepared with caution, in the same manner that one would behave in the clean room (i.e., opening tubes only when necessary).
14. Adding EtBr to the running buffer offers the most dilute option, but increases the potential for splashing and also leads to EtBr contamination of the gel apparatus. Addition of EtBr to the gel prior to setting also contaminates the apparatus, but keeps the EtBr more contained than in the buffer. However, if EtBr is added to the gel, it must *never* be added before heating, as EtBr will be aerosolized, which is hazardous. Staining the gel after electrophoresis requires a more concentrated solution of EtBr than would be added to the running buffer, but provides a very contained region of contamination and often produces sharper DNA band images. Always dispose of EtBr-contaminated waste following your institution's health and safety protocols.
15. Be sure that the samples are run slowly so that adequate separation occurs. As fragments are very small, it is important to obtain clear differentiation from primer-dimers, which are often not much smaller than the desired product.
16. If not all fragments are successfully amplified, the same two-step multiplex procedure can be applied including only the primers for the missing fragments, or, in rare cases, single PCRs must be performed to achieve amplification of all fragments.

Acknowledgments

The Pennsylvania State University and the National Science Foundation Award ANS-0909456 supported this work.

References

1. Hofreiter M, Jaenicke V, Serre D, Haeseler Av A, Pääbo S (2001) DNA sequences from multiple amplifications reveal artifacts induced by cytosine deamination in ancient DNA. Nucleic Acids Res 29:4793–4799
2. Chamberlain JS, Gibbs RA, Ranier JE, Nguyen PN, Caskey CT (1988) Deletion screening of the Duchenne muscular dystrophy locus via multiplex DNA amplification. Nucleic Acids Res 16:11141–11156
3. Edwards MC, Gibbs RA (1994) Multiplex PCR: advantages, development, and applications. PCR Methods Appl 3:65–75
4. Hummel S, Schultes T, Bramanti B, Herrmann B (1999) Ancient DNA profiling by megaplex amplications. Electrophoresis 20:1717–1721

5. Schultes T, Hummel S, Herrmann B (1999) Amplification of Y-chromosomal STRs from ancient skeletal material. Hum Genet 104:164–166
6. Krause J, Dear PH, Pollack JL, Slatkin M, Spriggs H, Barnes I, Lister AM, Ebersberger I, Pääbo S, Hofreiter M (2006) Multiplex amplification of the mammoth mitochondrial genome and the evolution of Elephantidae. Nature 439:724–727
7. Römpler H, Dear PH, Krause J, Meyer M, Rohland N, Schöneberg T, Spriggs H, Stiller M, Hofreiter M (2006) Multiplex amplification of ancient DNA. Nat Protoc 1:720–728
8. Rohland N, Malaspinas AS, Pollack JL, Slatkin M, Matheus P, Hofreiter M (2008) Proboscidean mitogenomics: chronology and mode of elephant evolution using mastodon as outgroup. PLoS Biol 5:e207
9. Richards MP, Pacher M, Stiller M, Quilès J, Hofreiter M, Constantin S, Zilhão J, Trinkaus E (2008) Isotopic evidence for omnivory among European cave bears: Late Pleistocene *Ursus spelaeus* from the Peştera cu Oase, Romania. Proc Natl Acad Sci USA 105:600–604
10. Krause J, Unger T, Noçon A, Malaspinas AS, Kolokotronis SO, Stiller M, Soibelzon L, Spriggs H, Dear PH, Briggs AW, Bray SC, O'Brien SJ, Rabeder G, Matheus P, Cooper A, Slatkin M, Pääbo S, Hofreiter M (2008) Mitochondrial genomes reveal an explosive radiation of extinct and extant bears near the Miocene-Pliocene boundary. BMC Evol Biol 8:220
11. Knapp M, Rohland N, Weinstock J, Baryshnikov G, Sher A, Nagel D, Rabeder G, Pinhasi R, Schmidt HA, Hofreiter M (2009) First DNA sequences from Asian cave bear fossils reveal deep divergences and complex phylogeographic patterns. Mol Ecol 18: 1225–1238
12. Stiller M, Baryshnikov G, Bocherens H, Grandal d'Anglade A, Hilpert B, Münzel SC, Pinhasi R, Rabeder G, Rosendahl W, Trinkaus E, Hofreiter M, Knapp M (2010) Withering away—25,000 years of genetic decline preceded cave bear extinction. Mol Biol Evol 27:975–978
13. Prost S, Knapp M, Flemmig J, Hufthammer AK, Kosintsev P, Stiller M, Hofreiter M (2010) A phantom extinction? New insights into extinction dynamics of the Don-hare *Lepus tanaiticus*. J Evol Biol 23:2022–2029
14. Römpler H, Rohland N, Lalueza-Fox C, Willerslev E, Kuznetsova T, Rabeder G, Bertranpetit J, Schöneberg T, Hofreiter M (2006) Nuclear gene indicates coat-color polymorphism in mammoths. Science 313:62
15. Campbell KL, Roberts JE, Watson LN, Stetefeld J, Sloan AM, Signore AV, Howatt JW, Tame JR, Rohland N, Shen TJ, Austin JJ, Hofreiter M, Ho C, Weber RE, Cooper A (2010) Substitutions in woolly mammoth hemoglobin confer biochemical properties adaptive for cold tolerance. Nat Genet 42:536–540
16. Stiller M, Knapp M, Stenzel U, Hofreiter M, Meyer M (2009) Direct multiplex sequencing (DMPS)—a novel method for targeted high-throughput sequencing of ancient and highly degraded DNA. Genome Res 19: 1843–1848

Chapter 18

Preparation of Next-Generation Sequencing Libraries from Damaged DNA

Adrian W. Briggs and Patricia Heyn

Abstract

Next-generation sequencing (NGS) has revolutionized ancient DNA research, especially when combined with high-throughput target enrichment methods. However, attaining high sequencing depth and accuracy from samples often remains problematic due to the damaged state of ancient DNA, in particular the extremely low copy number of ancient DNA and the abundance of uracil residues derived from cytosine deamination that lead to miscoding errors. It is therefore critical to use a highly efficient procedure for conversion of a raw DNA extract into an adaptor-ligated sequencing library, and equally important to reduce errors from uracil residues. We present a protocol for NGS library preparation that allows highly efficient conversion of DNA fragments into an adaptor-ligated form. The protocol incorporates an option to remove the vast majority of uracil miscoding lesions as part of the library preparation process. The procedure requires only two spin column purification steps and no gel purification or bead handling. Starting from an aliquot of DNA extract, a finished, highly amplified library can be generated in 5 h, or under 3 h if uracil removal is not required.

Key words: Ancient DNA, Ligation, Library preparation, Next-generation sequencing, High throughput sequencing, Damage repair

1. Introduction

The development of next-generation sequencing (NGS) techniques has revolutionized genomics. By avoiding the capillary electrophoresis that limits the throughput of traditional Sanger sequencing, highly miniaturized and parallelized NGS platforms can perform tens of millions of sequence reactions per machine run, making it possible to generate whole eukaryote genome sequences in a few days (1). The sequencing mechanisms and chemistries of current high-throughput platforms (Solexa-Illumina, 454-Roche, SOLiD-Applied Biosystems) are discussed elsewhere (2), but all current high-throughput platforms use essentially the

Beth Shapiro and Michael Hofreiter (eds.), *Ancient DNA: Methods and Protocols*, Methods in Molecular Biology, vol. 840,
DOI 10.1007/978-1-61779-516-9_18, © Springer Science+Business Media, LLC 2012

same sample preparation principle. This principle is the construction of a sequencing library, where all DNA fragments in a sample of interest are first end-repaired and then ligated to universal sequencing-adaptors. The universal adaptors allow amplification and sequencing of all the DNA fragments in parallel using a single primer pair and sequencing primer.

In addition to the high raw sequence output, sequencing from adaptor-ligated libraries is extremely well suited to ancient DNA (aDNA) work for at least five additional reasons: (1) the main drawback of NGS, short read lengths, is not a problem for aDNA since the DNA is already highly fragmented; (2) the use of adaptor primers outside the ancient fragments permits recovery of sequence information from molecules too short for conventional PCR, which is important since such fragments can constitute the vast majority of material in an ancient sample; (3) universal adaptors allow bulk PCR amplification of an entire library before any downstream experiments, effectively immortalizing the library and greatly reducing the material demands on irreplaceable biological samples (3, 4); (4) the 5' and 3' ends of fragments indicate unique DNA breakpoints and thus allow unique starting molecules to be counted at every genomic position (5), greatly increasing reliability and robustness against errors due to DNA damage or contamination; (5) library adaptors can be project-specifically barcoded, allowing downstream experiments to be performed outside the constrained environment of the aDNA cleanroom with no risk of fresh contamination by present day genomic DNA (6). Such downstream experiments often include targeted capture of genomic regions of interest prior to sequencing (4, 7, 8). Targeted capture avoids the high costs of sequencing irrelevant DNA from the sample, in particular the large amounts of microbial DNA present in most ancient remains.

The numerous advantages of aDNA sequencing by NGS approaches have allowed for the first time whole nuclear genomes to be sequenced from several thousand-year-old remains (9–12), as well as long target regions from multiple individuals for population genetic analyses (4, 13, 14). Despite these successes, however, challenges remain when applying NGS to aDNA. After the death of an organism, its DNA is degraded by endogenous nucleases. DNA preserved over long time periods is also damaged by chemical and physical events, causing strand-breaks and base-modifications. Thus, aDNA is invariably highly fragmented and damaged (15). Although NGS libraries can be indefinitely amplified by adaptor-primed PCR, a major limiting factor in the NGS approach is the efficiency with which aDNA molecules in the raw extract are successfully end-repaired and ligated to the adaptors during library preparation. Clearly an inefficient library preparation protocol will strongly counteract the advantages of an NGS approach. In addition, base-modifications are not repaired in standard NGS library preparation and can cause two problems: miscoding lesions and blocking lesions. The major source for miscoding lesions in aDNA is deamination of cytosine to uracil, which gives rise to C–T transitions (16). It has been shown that uracil deamination is greatly increased in the regions close to the ends of fragments (6), which are not analyzed in conventional PCR but form a substantial fraction of NGS data. Base modifications that block DNA polymerase during amplification remain poorly understood. However, recent evidence suggests that blocking lesions, if present, are at rather low frequency in aDNA and so may not be a major source of concern (17).

Here we describe a NGS library preparation protocol (Fig. 1) adapted to the requirements of aDNA work. In order to produce maximum library yield, the protocol employs an efficient ligation procedure, involves only two spin column purification steps before library amplification, and does not require any bead handling steps. An optional feature of the method is the inclusion of two repair enzymes (available pre-mixed from NEB), uracil-DNA-glycosylase and endonuclease VIII, that remove uracil and repair the DNA fragments afterwards, leaving the fragments amenable to sequencing yet free of miscoding lesions. This activity drastically reduces the transition error rate in final sequences (at least 50-fold (18)), and therefore under most circumstances we strongly recommend using these enzymes. However, for certain applications the retention of miscoding errors in final aDNA sequences may be desired, for example in studies of aDNA damage, or if damage itself will be used to assess the authenticity of presumed aDNA (19, 20). In these cases the enzymes can be left out.

Our protocol allows conversion of a raw aDNA extract into a PCR-amplified library in three to five hours. The procedure is fully compatible with Roche/454, Illumina/Solexa or other NGS sequencing platforms, simply requiring the appropriate sequences for the universal adaptors.

2. Materials

2.1. End Repair

1. 10× NEBuffer 2 (NEB, Ipswich, MA).
2. T4 polynucleotide kinase, 10 units/μL (NEB, Ipswich, MA).
3. T4 DNA polymerase, 3 units/μL (NEB, Ipswich, MA).
4. USER enzyme, 1 unit/μL (NEB, Ipswich, MA).

2.2. Adaptor Attachment

1. Double-stranded library adaptors (2.5 mM) (see Note 1).
2. Quick ligation kit (NEB, Ipswich, MA).
3. 3.dNTP mix (25 mM each of dATP, dTTP, dCTP, dGTP).

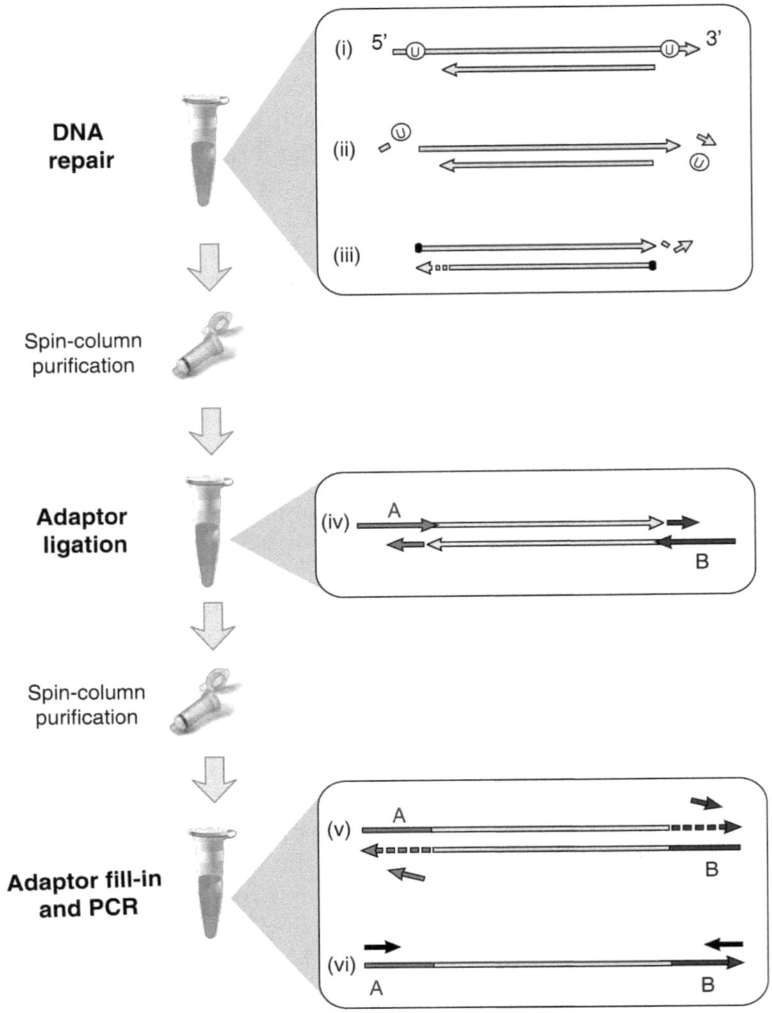

Fig. 1. The library preparation procedure. (i) Damage in ancient DNA leaves 5' and 3' overhangs and uracil (U) bases; (ii) USER enzyme removes and repairs uracil sites; (iii) *T4* PNK polymerase and *T4* polymerase phosphorylate 5' ends (*black*) and repair overhangs; (iv) *T4* ligase attaches nonphosphorylated A and B adaptors to the fragments at one end only; (v) *Bst* polymerase fills in from the nick to complete adaptor attachment; (vi) Adaptor-complementary primers and *Taq* polymerase amplify the entire library to immortalize the sample ahead of direct sequencing or capture and sequencing.

4. 10× Thermopol buffer (NEB, Ipswich, MA).
5. *Bst* DNA polymerase large fragment, 8 units/μL (NEB, Ipswich, MA).

2.3. Primary Library Amplification

1. 10× Thermopol buffer (NEB, Ipswich, MA).
2. AmpliTaq Gold DNA Polymerase, 5 units/μL (Applied Biosystems, Foster City, CA).
3. dNTP mix (25 mM each of dATP, dTTP, dCTP, dGTP).

4. Adaptor-specific PCR primer pair (recommended annealing temperature 60°C).
5. Siliconized 1.7-mL tubes (Sigma-Aldrich, St. Louis, MO, cat T3406-250EA).

2.4. Secondary Library Amplification

1. Phusion High-Fidelity PCR master mix with HF buffer (Finnzymes/NEB, Ipswich, MA).
2. Adaptor-specific PCR primer pair.
3. Agilent DNA 1000 chip.
4. Agilent 2100 Bioanalyzer.

2.5. Library Quantification

1. DyNAmo™ Flash SYBR® Green qPCR Kit (Finnzymes/NEB, Ipswich, MA).
2. Adaptor-specific PCR primer pair.
3. TE buffer (10-mM Tris–HCl pH 8.0, 1-mM EDTA).
4. Tween-20.
5. qPCR machine.

All steps

1. MinElute PCR purification kit (Qiagen, Hilden, Germany).
2. Molecular grade water.
3. Thermal cycler.

3. Methods

All aDNA works must be done in a dedicated cleanroom. Once the molecules are ligated to suitable adaptors the library can be handled outside the cleanroom. Still, one must be careful to avoid cross-library contamination, especially when handling PCR-amplified libraries.

3.1. aDNA Repair

Damaged DNA such as aDNA can contain overhanging 5' or 3' ends and deaminated cytosines in the form of uracil bases. Two alternative repair reactions, A and B, are presented below. Reaction A excises uracil bases from the DNA, cleaves the resulting abasic sites, and repairs remaining overhangs to leave ligatable blunt ends. Reaction B repairs overhanging ends but does not excise uracils. Choose one of these reactions depending on the results desired.

Reaction A: End repair with uracil removal

1A. Prepare a reaction mix containing 5-μL 10× NEBuffer 2, 23-μL aDNA extract, and 17-μL molecular grade water. Mix reaction and briefly centrifuge to collect the reagents at the bottom of the tube (see Notes 2–5).

2A. Add 2-μL T4 polynucleotide kinase and 3-μL USER enzyme (see Note 6). Mix by gently flicking the tube and briefly centrifuge to collect the reagents at the bottom. Incubate for 3 h at 37°C in a thermal cycler.

3A. Add 1-μL T4 DNA polymerase to the reaction, mix by gently flicking the tube, and centrifuge briefly. Incubate at 25°C for 30 min in a thermal cycler. Proceed to step 1 in Subheading 3.1.

Reaction B: End repair without uracil removal

1B. Prepare a reaction mix containing 5-μL 10× NEBuffer 2, 23-μL aDNA extract (see Notes 2–5), and 19-μL molecular grade water. Mix reaction and briefly centrifuge to collect the reagents at the bottom of the tube.

2B. Add 1-μL T4 polynucleotide kinase and 2-μL T4 DNA polymerase. Mix by gently flicking the tube and briefly centrifuge to collect the reagents at the bottom. Incubate for 30 min at 25°C in a thermal cycler. Proceed to step 1 in Subheading 3.1.

1. *(For both repair reactions A and B)* Purify the reaction with a MinElute spin column. Load column by adding 150-μL buffer PB to the reaction, mix carefully by pipetting up and down several times, then transfer to the spin column. After binding and washing according to the manufacturer's instructions, elute DNA from the column in 15-μL buffer EB. After adding buffer EB to the column incubate for 1–2 min before centrifugation, to improve DNA recovery.

3.2. Adaptor Attachment

Having repaired the DNA fragments, this step attaches universal adaptor sequences to the fragments in two steps; first, the 5' ends of the fragments are ligated to 3' ends of the adaptors; second, the 3' ends of the fragments are extended along the ligated adaptor strands to complete double-stranded adaptor attachment (see (iv) and (v) in Fig. 1). We recommend including an extraction negative (blank) and a blank library control (water only) in the library preparation for quality control.

1. Prepare a reaction mix containing 20-μL 2× Quick ligase buffer, 1-μL 2.5 μM adaptors, 15-μL purified repaired DNA, and 3-μL molecular grade water. Mix all components well by pipetting up and down several times. After mixing the components, add 1-μL Quick ligase. Mix by flicking thoroughly, and spin down briefly to bring the mixture to the bottom of the tube (see Note 7).
2. Incubate the ligation reaction for 5 min at 25°C in a thermal cycler. Purify with a Qiagen MinElute column. Load column by adding 160-μL buffer PB to the ligation product, mixing thoroughly by pipetting, and transferring to the spin column.

After binding and washing according to the manufacturer's instructions, elute the DNA from the column with 15-μL EB. After adding buffer EB to the column, incubate for 1–2 min before centrifugation to improve DNA recovery.

3. Prepare a reaction mix containing 5-μL 10× Thermopol buffer, 0.5-μL dNTP mix (25-mM each dNTP) 15-μL ligated and purified DNA, and 27.5-μL molecular grade water. Mix the reaction and briefly centrifuge down. Add 2-μL *Bst* DNA polymerase, mix by gently flicking the tube and briefly centrifuge down. Incubate at 37°C for 30 min in a thermal cycler, followed by 20 min at 80°C to inactivate the *Bst* polymerase. Remove a 2 μL aliquot of the product for library quantification with qPCR (for qPCR instructions, see Subheading 3.5) (see Note 8).

3.3. Primary Library Amplification

In this step, the entire library is PCR-amplified using the universal priming sites provided by the library adaptors. The reaction takes place in the same tube as adaptor fill-in without purification, avoiding any loss of material (see Note 9). Immediate amplification effectively immortalizes the library by making many copies of every original template molecule.

1. Prepare a 50 μL volume of "PCR addition mix" containing 5-μL 10× Thermopol buffer, 0.5-μL dNTP mix (25 mM each dNTP), 5 μL each adaptor primer (10-μL M stock), and 1-μL AmpliTaq Gold DNA Polymerase (see Note 10). Mix all 50 μL into the heat-inactivated 50 μL fill-in reaction. Transfer to a PCR machine and perform the following cycling conditions:

95°C	12 min		
95°C	30 s	|	
60°C	30 s	|	×12
72°C	1 min	|	
72°C	3 min		

2. Purify the amplified product with a Qiagen MinElute spin column. Elute in 50-μL EB (see Note 11). Quantify the product with qPCR alongside the retained aliquot of unamplified library, to check for amplification success (see Note 12).

3.4. Secondary Library Amplification

For most applications, higher-fold amplification of the library will be necessary, requiring a second library PCR. For example, array hybridization capture requires up to 20 μg of amplified library product (7). For secondary amplifications, we recommend using a proofreading polymerase such as the Phusion enzyme described here, to reduce the occurrence of PCR errors in final sequences.

1. Prepare a PCR at a total volume 100 μL containing 50-μL 2× Phusion HF master mix (Finnzymes/NEB, Ipswich, MA), 5 μL each adaptor primer (10-μM stock), and 10 μL of primary-amplified library.
2. Transfer to a PCR block or thermal cycler and perform the following cycling conditions:

98°C	1 min	
98°C	30 s	
60°C	30 s	×10–20 depending on desired yield (see Note 13)
72°C	1 min	
72°C	3 min	

3. Purify the amplified product with a Qiagen MinElute spin column. Elute in 50-μL EB. Quantify the product with Nanodrop (less accurate) or an Agilent Bioanalyzer 2100 DNA 1000 chip (more accurate).

The repaired, immortalized, and quantified DNA library is now ready for direct high-throughput sequencing or target enrichment with capture methods such as array hybridization (7), in-solution hybridization capture (8), or primer extension capture (4) followed by sequencing.

3.5. Quantification of Library by qPCR

Conventional quantification of unamplified aDNA sequencing libraries is not generally possible by UV absorbance or fluorescence measurements due to the very low amounts of material (below detection limits). While it is not strictly necessary to quantify the unamplified library, as the secondary-amplified (and sometimes primary-amplified) library can be quantified by UV or fluorescence, it is usually very useful to know how much DNA the primary library contained, and whether the amplification steps worked efficiently. To quantify libraries of low concentration, Meyer et al. (6) developed a qPCR quantification method (21). Here we describe briefly an updated version of that protocol. For further information on qPCR in aDNA work see Chap. 16.

1. Prepare a dilution series of a sequencing library with known concentration, using 1× TE buffer to dilute. The range should be from 10^9 molecules/μL to 10^2 molecules/μL in tenfold dilution steps. This will be your standard for quantification (see Note 14).
2. Prepare a qPCR Mastermix by adding molecular biology grade water and adaptor-PCR primers to the DyNAmo™ Flash SYBR® Green qPCR Kit according to the manufacturer's instructions.

3. Dispense the mastermix into a qPCR-suitable plate.
4. Add 1 μL each template. Measure the standard dilution series at least twice. For quality control, include the unamplified library, primary-amplified library, secondary-amplified library (if relevant), unamplified extraction negative (blank) library, unamplified blank library, and qPCR-blank, all in duplicates or triplicates. It might be necessary to include dilutions from the amplified libraries (even up to 10,000-fold) to be in the quantification range of the standard.
5. Perform qPCR with the cycling conditions following the instruction of the manufacturer, using the appropriate annealing temperature for the primer pair used.
6. Analyze the qPCR according to instructions from the qPCR machine manufacturer (see Note 15).

4. Notes

1. The method we present here is compatible with any library adaptor sequences, although the oligonucleotide design may differ from manufacturers' own kits due to differences in the ligation strategy. To make library A and B adaptors suitable for our ligation procedure, the adaptors must be designed as follows: for each double-stranded adaptor, the oligo that attaches to the 5' end of the insert DNA should be ordered full-length, whereas the oligo that will attach to the 3' end of the insert DNA should be ordered truncated to 12–14 nt, as first described in (22). Prepare the adaptors from lyophilized oligos as follows: dissolve each constituent oligo to 100 μM in TE buffer; mix the two oligos of each adaptor 1:1, heat to 95°C for 10 s then allow to cool to RT; mix the two resulting double-stranded adaptors 1:1, and dilute in TE 1:10 to form a final 2.5-μM double-stranded adaptor stock.
2. Our library preparation procedure is compatible with aDNA extracted using any method, as long as DNA was not denatured during the extraction (adaptor ligation requires double-stranded molecules so denatured fragments cannot get into the library). Given the short lengths of many aDNA fragments, this means keeping the temperature during extraction below ~60°C at all times.
3. The raw extract must be stored in a low retention tube, such as a Sigma-Aldrich siliconized 1.7-mL tube. DNA stored in regular polypropylene tubes can decrease dramatically in concentration over time due to DNA sticking to tube walls.
4. As much extract should be used as possible for aDNA library preparation, as the quality of the final results depends on the

level of genomic coverage contained within the library. However, we have not tested the protocol using more than 23 μL of pure extract in the 50 μL repair reaction. If more than 23 μL of pure extract is to be used, we recommend making multiple parallel libraries and pooling them prior to the primary amplification reaction. However, if an extract will be difficult or impossible to replace, we recommend not using more than half of the extract in any single experiment, to keep some material behind in case something goes wrong.

5. Do not vortex mixtures containing enzymes as this may reduce enzyme activity.
6. USER enzyme is a convenient mixture of UDG and endoVIII sold by NEB. The exact ratio of the enzymes is proprietary, but if USER is not available, it can be replaced in the 50 μL repair reaction by 2 μL endoVIII (10 U/μL) and 1-μL UDG (5 U/μL) (NEB unit definitions).
7. It is critical to add the ligase only after the other components have been mixed in order to avoid contact between ligase and locally high concentrations of incompletely mixed DNA inserts or adaptors. This could lead to increased chimeric insert or adaptor dimer formation. Since the Quick ligase buffer contains PEG-6000, which is highly viscous, be aware that it can take slightly more pipetting/flicking than usual to thoroughly mix the ligation reagents.
8. *Bst* polymerase has a strong strand displacement function and can use the unligated 3’ OH of the DNA insert fragment as a primer for second strand synthesis of the adaptor. It will displace the short arm of the partially double-stranded adaptor and fill in the missing part (Fig. 1 (v)).
9. Unlike previous studies, this protocol performs the fill-in step in solution and does not use streptavidin beads, as we have found that beads are unnecessary. By removing this step, no material is lost between fill-in and library amplification, increasing yield and reproducibility and at the same time simplifying the protocol.
10. We recommend Amplitaq Gold DNA polymerase (Applied Biosystems) for the primary library amplification as we have found it to work well in the Thermopol buffer that is used for the fill-in polymerase.
11. All steps subsequent to adaptor ligation can be performed outside the aDNA cleanroom. Take care, however, to avoid library cross-contamination.
12. The yield of libraries prepared from aDNA can vary dramatically depending on the sample, from virtually zero to 10^{12} ligated insert molecules. In any case, however, 12 cycles of primary amplification should be enough to produce over 1,000

copies of each unique starting molecule unless the starting concentration is very high and PCR is saturated early in the exponential phase (in which case amplification is less necessary anyway).

13. A 100-μL Phusion PCR reaction of an aDNA sequencing library will produce a maximum of ~1 μg of amplified library DNA. If higher amounts of library are required, multiple parallel PCRs can be performed.
14. An accurate dilution series is crucial for quantification of your library. For convenience, we recommend storing the standards in eight-tube 0.2-mL PCR strips to allow multichannel pipetting into the qPCR reactions. If the strips are not siliconized/with low-retention, add 0.05% Tween-20 to the TE buffer to avoid DNA sticking to the tube walls over time.
15. The extraction negative (blank) library and negative library control (water only) will not be DNA-free, but should display only short adaptor-artifacts when visualized on a gel after PCR. A smear of longer molecules may indicate contamination. The concentration of molecules in these control libraries should be lower than in the sample library.

References

1. Kircher M, Kelso J (2010) High-throughput DNA sequencing—concepts and limitations. Bioessays 32:524–536
2. Metzker ML (2010) Sequencing technologies—the next generation. Nat Rev Genet 11:31–46
3. Blow MJ, Zhang T, Woyke T, Speller CF, Krivoshapkin A, Yang DY, Derevianko A, Rubin EM (2008) Identification of ancient remains through genomic sequencing. Genome Res 18:1347–1353
4. Briggs AW, Good JM, Green RE, Krause J, Maricic T, Stenzel U, Lalueza-Fox C, Rudan P, Brajkovic D, Kucan Z, Gusic I, Schmitz R, Doronichev VB, Golovanova LV, de la Rasilla M, Fortea J, Rosas A, Paabo S (2009) Targeted retrieval and analysis of five Neandertal mtDNA genomes. Science 325:318–321
5. Green RE, Krause J, Ptak SE, Briggs AW, Ronan MT, Simons JF, Du L, Egholm M, Rothberg JM, Paunovic M, Paabo S (2006) Analysis of one million base pairs of Neanderthal DNA. Nature 444:330–336
6. Briggs AW, Stenzel U, Johnson PL, Green RE, Kelso J, Prufer K, Meyer M, Krause J, Ronan MT, Lachmann M, Paabo S (2007) Patterns of damage in genomic DNA sequences from a Neandertal. Proc Natl Acad Sci USA 104: 14616–14621
7. Burbano HA, Hodges E, Green RE, Briggs AW, Krause J, Meyer M, Good JM, Maricic T, Johnson PL, Xuan Z, Rooks M, Bhattacharjee A, Brizuela L, Albert FW, de la Rasilla M, Fortea J, Rosas A, Lachmann M, Hannon GJ, Paabo S (2010) Targeted investigation of the Neandertal genome by array-based sequence capture. Science 328:723–725
8. Maricic T, Whitten M, Paabo S (2010) Multiplexed DNA sequence capture of mitochondrial genomes using PCR products. PLoS One 5:e14004
9. Green RE, Krause J, Briggs AW, Maricic T, Stenzel U, Kircher M, Patterson N, Li H, Zhai W, Fritz MH, Hansen NF, Durand EY, Malaspinas AS, Jensen JD, Marques-Bonet T, Alkan C, Prufer K, Meyer M, Burbano HA, Good JM, Schultz R, Aximu-Petri A, Butthof A, Hober B, Hoffner B, Siegemund M, Weihmann A, Nusbaum C, Lander ES, Russ C, Novod N, Affourtit J, Egholm M, Verna C, Rudan P, Brajkovic D, Kucan Z, Gusic I, Doronichev VB, Golovanova LV, Lalueza-Fox C, de la Rasilla M, Fortea J, Rosas A, Schmitz RW, Johnson PL, Eichler EE, Falush D, Birney E, Mullikin JC, Slatkin M, Nielsen R, Kelso J, Lachmann M, Reich D, Paabo S (2010) A draft sequence of the Neandertal genome. Science 328:710–722

10. Miller W, Drautz DI, Ratan A, Pusey B, Qi J, Lesk AM, Tomsho LP, Packard MD, Zhao F, Sher A, Tikhonov A, Raney B, Patterson N, Lindblad-Toh K, Lander ES, Knight JR, Irzyk GP, Fredrikson KM, Harkins TT, Sheridan S, Pringle T, Schuster SC (2008) Sequencing the nuclear genome of the extinct woolly mammoth. Nature 456:387–390
11. Rasmussen M, Li Y, Lindgreen S, Pedersen JS, Albrechtsen A, Moltke I, Metspalu M, Metspalu E, Kivisild T, Gupta R, Bertalan M, Nielsen K, Gilbert MT, Wang Y, Raghavan M, Campos PF, Kamp HM, Wilson AS, Gledhill A, Tridico S, Bunce M, Lorenzen ED, Binladen J, Guo X, Zhao J, Zhang X, Zhang H, Li Z, Chen M, Orlando L, Kristiansen K, Bak M, Tommerup N, Bendixen C, Pierre TL, Gronnow B, Meldgaard M, Andreasen C, Fedorova SA, Osipova LP, Higham TF, Ramsey CB, Hansen TV, Nielsen FC, Crawford MH, Brunak S, Sicheritz-Ponten T, Villems R, Nielsen R, Krogh A, Wang J, Willerslev E (2010) Ancient human genome sequence of an extinct Palaeo-Eskimo. Nature 463:757–762
12. Reich D, Green RE, Kircher M, Krause J, Patterson N, Durand EY, Viola B, Briggs AW, Stenzel U, Johnson PL, Maricic T, Good JM, Marques-Bonet T, Alkan C, Fu Q, Mallick S, Li H, Meyer M, Eichler EE, Stoneking M, Richards M, Talamo S, Shunkov MV, Derevianko AP, Hublin JJ, Kelso J, Slatkin M, Paabo S (2010) Genetic history of an archaic hominin group from Denisova Cave in Siberia. Nature 468:1053–1060
13. Gilbert MT, Drautz DI, Lesk AM, Ho SY, Qi J, Ratan A, Hsu CH, Sher A, Dalen L, Gotherstrom A, Tomsho LP, Rendulic S, Packard M, Campos PF, Kuznetsova TV, Shidlovskiy F, Tikhonov A, Willerslev E, Iacumin P, Buigues B, Ericson PG, Germonpre M, Kosintsev P, Nikolaev V, Nowak-Kemp M, Knight JR, Irzyk GP, Perbost CS, Fredrikson KM, Harkins TT, Sheridan S, Miller W, Schuster SC (2008) Intraspecific phylogenetic analysis of Siberian woolly mammoths using complete mitochondrial genomes. Proc Natl Acad Sci U S A 105:8327–8332
14. Stiller M, Knapp M, Stenzel U, Hofreiter M, Meyer M (2009) Direct multiplex sequencing (DMPS)—a novel method for targeted high-throughput sequencing of ancient and highly degraded DNA. Genome Res 19:1843–1848
15. Hofreiter M, Serre D, Poinar HN, Kuch M, Pääbo S (2001) Ancient DNA. Nat Rev Genet 2:353–359
16. Hofreiter M, Jaenicke V, Serre D, Haeseler Av A, Paabo S (2001) DNA sequences from multiple amplifications reveal artifacts induced by cytosine deamination in ancient DNA. Nucleic Acids Res 29:4793–4799
17. Heyn P, Stenzel U, Briggs AW, Kircher M, Hofreiter M, Meyer M (2010) Road blocks on paleogenomes—polymerase extension profiling reveals the frequency of blocking lesions in ancient DNA. Nucleic Acids Res 38:e161
18. Briggs AW, Stenzel U, Meyer M, Krause J, Kircher M, Paabo S (2010) Removal of deaminated cytosines and detection of in vivo methylation in ancient DNA. Nucleic Acids Res 38:e87
19. Krause J, Briggs AW, Kircher M, Maricic T, Zwyns N, Derevianko A, Paabo S (2010) A complete mtDNA genome of an early modern human from Kostenki, Russia. Curr Biol 20:231–236
20. Krause J, Fu Q, Good JM, Viola B, Shunkov MV, Derevianko AP, Paabo S (2010) The complete mitochondrial DNA genome of an unknown hominin from southern Siberia. Nature 464:894–897
21. Meyer M, Briggs AW, Maricic T, Hober B, Hoffner B, Krause J, Weihmann A, Paabo S, Hofreiter M (2008) From micrograms to picograms: quantitative PCR reduces the material demands of high-throughput sequencing. Nucleic Acids Res 36:e5
22. Margulies M, Egholm M, Altman WE, Attiya S, Bader JS, Bemben LA, Berka J, Braverman MS, Chen YJ, Chen Z, Dewell SB, Du L, Fierro JM, Gomes XV, Godwin BC, He W, Helgesen S, Ho CH, Irzyk GP, Jando SC, Alenquer ML, Jarvie TP, Jirage KB, Kim JB, Knight JR, Lanza JR, Leamon JH, Lefkowitz SM, Lei M, Li J, Lohman KL, Lu H, Makhijani VB, McDade KE, McKenna MP, Myers EW, Nickerson E, Nobile JR, Plant R, Puc BP, Ronan MT, Roth GT, Sarkis GJ, Simons JF, Simpson JW, Srinivasan M, Tartaro KR, Tomasz A, Vogt KA, Volkmer GA, Wang SH, Wang Y, Weiner MP, Yu P, Begley RF, Rothberg JM (2005) Genome sequencing in microfabricated high-density picolitre reactors. Nature 437: 376–380

Chapter 19

Generating Barcoded Libraries for Multiplex High-Throughput Sequencing

Michael Knapp, Mathias Stiller, and Matthias Meyer

Abstract

Molecular barcoding is an essential tool to use the high throughput of next generation sequencing platforms optimally in studies involving more than one sample. Various barcoding strategies allow for the incorporation of short recognition sequences (barcodes) into sequencing libraries, either by ligation or polymerase chain reaction (PCR). Here, we present two approaches optimized for generating barcoded sequencing libraries from low copy number extracts and amplification products typical of ancient DNA studies.

Key words: Next generation high-throughput sequencing, Direct multiplex sequencing, Multiplex PCR, DNA capture, Ancient DNA, Barcoding

1. Introduction

1.1. Background

High-throughput sequencing produces enormous amounts of sequence data compared to traditional Sanger sequencing (1). For many studies, even the smallest lane on any next generation sequencing (NGS) instrument will produce excessive amounts of sequence data for a single sample. Moreover, if larger numbers of samples are to be analyzed, the cost soon becomes prohibitive if a full single lane is used per sample. It is therefore important to be able to pool multiple samples and sequence them in a single lane. As the information about sample origin of individual sequence reads is lost in all NGS approaches, this requires barcoding techniques, in which a specific tag is attached to all DNA fragments allowing them to be sorted bioinformatically after sequencing (2).

The most efficient way to produce a barcoded sequencing library is to amplify a genomic target region using polymerase chain reaction (PCR) with target-specific primers that include a

Beth Shapiro and Michael Hofreiter (eds.), *Ancient DNA: Methods and Protocols*, Methods in Molecular Biology, vol. 840, DOI 10.1007/978-1-61779-516-9_19, © Springer Science+Business Media, LLC 2012

sequencing adapter and barcode tail. While this approach is fast and simple, it has some drawbacks that limit its use. A complete barcoded set of primers is needed for each sample that is to be sequenced, which makes this strategy expensive for population level studies. It is also not suitable to barcode shotgun sequencing libraries.

Most other barcoding strategies rely on ligation of barcodes or barcoded sequencing adapters to amplified or non-amplified target DNA molecules. While more template DNA is lost than in a barcoding protocol using tailed PCR primers, this approach is suitable for all double-stranded DNA. The manufacturers of all major NGS platforms provide protocols for adapter ligation and further protocols are available in the scientific literature (e.g., (3, 4)). Most of these, however, are not designed for ancient DNA applications. Here, we present two protocols that were developed with specific ancient DNA applications in mind. Protocol 1 was optimized for producing barcoded sequencing libraries from highly degraded, low copy number DNA extracts. Protocol 2 was designed for barcoding preamplified multiplex PCR products, but can also be used for barcoding regular PCR products. Both protocols are derived from the original 454 library preparation protocol by Margulies et al. (5). As most ancient DNA NGS studies to date have used either the Roche 454 or the Illumina Solexa sequencing platform, we will focus on barcoding strategies for these instruments. Prior to introducing these protocols, we first present some brief recommendations about designing the adapters that will be used to barcode the sequences.

1.2. Adapter Design

As more than 99% of template molecules can be lost in the adapter ligation process (6) barcoded sequencing libraries of ancient DNA extracts should be obtained after no more than one adapter ligation. Thus protocols such as Parallel Tagged Sequencing (PTS) (3) that ligate the barcodes and sequencing adapters independently are not ideal for barcoding of low copy number templates. Consequently, barcodes must be designed to contain the complete sequencing adapter or a portion of the sequencing adapter for the sequencing platforms that is to be used. Protocol 1 requires the adapters to be biotinylated for a post-ligation purification step (steps 7–10). However, downstream applications such as Primer Extension Capture (PEC) (7) require nonbiotinylated libraries. We therefore recommend ligating truncated, biotinylated, and barcoded adapters to the target initially, and subsequently bring them to full length by amplification with tailed primers containing the remaining adapter sequence (Fig. 1). This has the additional benefit of reducing the length of the adapters and thereby the cost of the oligonucleotides. In general, one barcode per molecule is sufficient. Using barcodes on both sides of the target molecule (double barcoding)

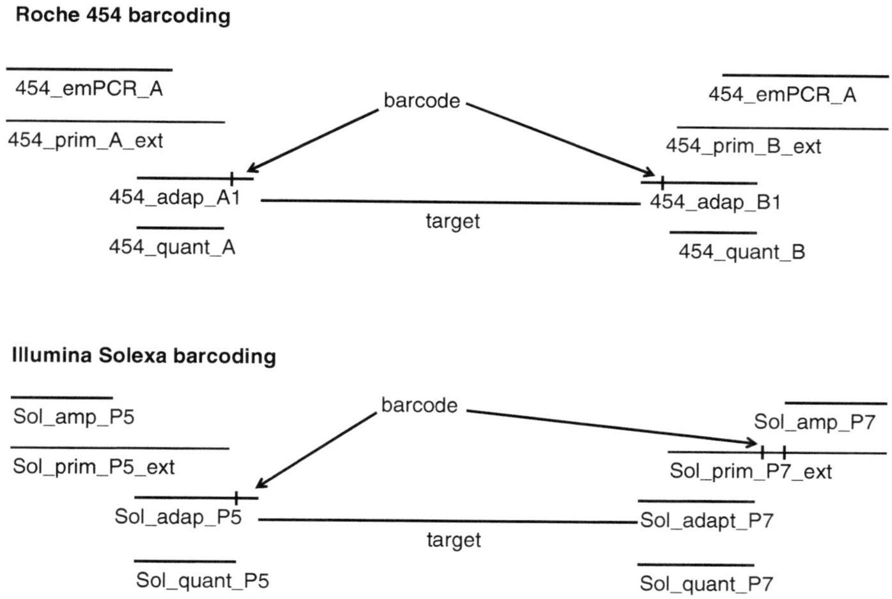

Fig. 1. Scheme of relative location of adapters, primers, and barcodes for Roche's 454 Titanium and Illumina's Solexa platform, respectively. Note that the barcodes on the A adapter (454) and P5 adapter (Illumina) are optional.

in most cases reduces the target read length, but provides valuable additional information to assess the quality of the data. For example, double barcoding makes it possible to identify cross-contamination between barcodes and potential jumping PCR artifacts (8).

Table 1 provides example adapter sequences for the Roche 454 sequencing platform and Table 2 shows example adapter sequences for the Illumina Solexa sequencing platform. For both sequencing platforms, each adapter consists of two single-stranded oligos that need to be hybridized to form the double-stranded adapter. To achieve directional blunt-end ligation, the plus strand oligo should be longer than its complement with the overhang on the 5′ side of the plus strand (i.e., the side that does not ligate to the target). To achieve better ligation efficiency and adapter stability, we recommend using phosphorothioate (PTO) bonds for the four 5′ most and the four 3′ most nucleotides of each oligo (9).

Illumina's indexing system differs in that the barcodes (indexes) are placed *within* one of the adapters (P7) rather than at the end of the adapter (3, 10) (Fig. 1). The barcode sequence is identified in a separate short sequencing read. This setup allows for a high degree of flexibility in experimental design, because libraries are first prepared with universal adapters, and different indexes can be added repeatedly by amplification with tailed primers just prior to target capture or sequencing (4). However, if double barcoding is to be used without modifying the sequencing software, the second

Table 1
Example for designing barcoded 454 titanium shotgun adapters

Laboratory code (Name in protocol)	Oligo sequence and description
454_adap_A1 (Adapter_1)	T*G*C*G*TGTCTCCGACTCAGacac*a*c*a*c Truncated A-Adapter with barcode (small letters are barcode)
454_adap_A1_rev (Adapter_1_rev)	g*t*g*t*gtgt*C*T*G*A Complement of barcoded A-Adapter (small letters are barcode)
454_adap_B1 (Adapter_2)	[Btn]T*G*C*C*TTGGCAGTCTCAGgtgt*g*t*g*t Truncated, biotinylated B-Adapter with barcode (small letters are barcode) *Barcode optional* *No biotin required for Protocol 2*
454_adap_B1_rev (Adapter_2_rev)	a*c*a*c*acac*C*T*G*A Complement of biotinylated B-Adapter with barcode (small letters are barcode) *Barcode optional*
454_prim_A_ext (ext_primer_F)	*CCATCTCATCCC*TGCGTGTCTCCGACTCAG Tailed A primer to extend truncated A-Adapter to full length (Tail in italics)
454_prim_B_ext (ext_primer_R)	*CCTATCCCCTGTG*TGCCTTGGCAGTCTCAG Tailed B primer to extend truncated B-Adapter to full length (Tail in italics)
454_quant_A (quant_primer_F1)	TGCGTGTCTCCGACTCAG Short primer to amplify and quantify library with truncated adapters in quantitative PCR
454_quant_B (quant_primer_R1)	TGCCTTGGCAGTCTCAG Short primer to amplify and quantify library with truncated adapters in quantitative PCR
454_emPCR_A (Amp_primer_F)	CCATCTCATCCCTGCGTGTC Short primer to amplify and quantify library with complete adapters (after extension with 454_prim_A_ext) in quantitative PCR
454_emPCR_B (Amp_primer_R)	CCTATCCCCTGTGTGCCTTG Short primer to amplify and quantify library with complete adapters (after extension with 454_prim_B_ext) in quantitative PCR

*Phosphorothioate (PTO) bond
[Btn]: Biotin; A list of suitable barcodes can for example be found in (3)

barcode still has to be attached to the truncated P5 adapter in the same way as for the 454 platform. The truncated P7 adapter does not have to be barcoded as the barcode is part of the tailed primer ("Sol_prim_P7_ext" in Fig. 1 and Table 2) used to extend the truncated P7 adapter to full length.

Table 2
Example for designing barcoded Illumina Solexa adapters

Laboratory code (Name in protocol)	Oligo sequence and description
Sol_adap_P5_1 (Adapter_1)	A*C*A*C*TCTTTCCCTACACGACGCTCTTCCGATCTac*a*g*t*g Truncated P5-Adapter with barcode (small letters are barcode) *barcode optional*
Sol_adap_P5_1_rev (Adapter_1_rev)	c*a*c*t*gtAGATCGGA*A*G*A*G Complement of barcoded P5-Adapter (small letters are barcode) *barcode optional*
Sol_adap_P7 (Adapter_2)	[Btn] G*T*G*A*CTGGAGTTCAGACGTGTGCTCTTCCG*A*T*C*T Truncated, biotinylated P7-Adapter *No biotin required for Protocol 2*
Sol_adap_P7_rev (Adapter_2_rev)	A*G*A*T*CGGA*A*G*A*G Complement of biotinylated P7-Adapter
Sol_prim_P5_ext (ext_primer_F)	*AATGATACGGCGACCACCGAGATCTA*CACTCTTTCCCTACACGA CGCTCTT Tailed P5 primer to extend truncated P5-Adapter to full length (tail in italics)
Sol_prim_P7_ext_1 (ext_primer_R)	*CAAGCAGAAGACGGCATACGAGAT*cactgtGTGACTGGAGTTC AGACGTGT Tailed P7 primer to extend and barcode truncated P7-Adapter (tail in italics, small letters are barcode)
*Sol_amp_P5 (Amp_primer_F)	AATGATACGGCGACCACCGA Full-length-library amplification primer P5 (for amplification of barcoded, preamplified)
*Sol_amp_P7 (Amp_primer_R)	CAAGCAGAAGACGGCATACGA Full-length-library amplification primer P7 (for amplification of barcoded, preamplified libraries)
Sol_quant_P5 (quant_primer_F)	ACACTCTTTCCCTACACGACGCTCTT Short primer to amplify and quantify truncated library in quantitative PCR
Sol_quant_P7 (quant_primer_R)	GTGACTGGAGTTCAGACGTGT Short primer to amplify and quantify truncated library in quantitative PCR

[Btn]: Biotin
*PTO bond. *As the tailed P7 primer also includes a barcode, they cannot be used as universal primers to amplify differently barcoded libraries. It is therefore recommended to use these truncated primers to amplify barcoded libraries. A list of suitable barcodes can for example be found in (4)

2. Materials

2.1. Materials Needed for Protocol 1 and Protocol 2

1. T4 DNA Polymerase.
2. T4 Polynucleotide Kinase.
3. T4 Ligase, including 50% PEG-4000 solution.

4. Bst DNA Polymerase, Large Fragment, including 10× ThermoPol buffer.
5. 10× Buffer Tango.
6. ATP, 100 mM stock solution.
7. dNTPs, 25 mM each.
8. Water, ultrapure.
9. 0.1× TE buffer (with and without 0.05% Tween 20): 10 mM Tris–HCl, 0.1 mM EDTA, pH 8.0. To prepare 50 mL of 0.1× TE, add 10 μL 0.5 M EDTA, pH 8.0 and 500 μL 1 M Tris, pH 8.0 to a 50-mL falcon tube and add ultrapure water to 50 mL. If desired, add 25 μL Tween 20 to obtain 0.1× TET (0.05% Tween 20) (see Note 1).
10. Qiagen MinElute PCR Purification Kit (for Protocol 2 this Purification kit is only needed if amplification products shorter than 100 bp are to be purified).
11. 10× Oligo hybridization buffer: 500 mM NaCl, 10 mM Tris–Cl, 1 mM EDTA, pH 8.0. Fill 5 mL of 5 M NaCl into a falcon tube and add 45 mL of 1× TE.
12. Agencourt AMPure XP DNA purification kit (Beckman Coulter) (Solid Phase Reversible Immobilization, (SPRI) Technology).
13. 70% Ethanol.
14. Amplitaq Gold DNA Polymerase with Buffer II and $MgCl_2$ solution (Applied Biosystems).
15. Barcoded adapters and amplification primers (as above).
16. Agencourt SPRICourt 96R Magnet Plate (Beckman Coulter) or 96 well Magnetic-Ring Stand (Applied Biosystems).
17. Thermal cycler with heated lid.
18. Real-time PCR machine.
19. Quantitative PCR reagents suitable for the real-time PCR machine use. For Protocol 1, it is essential that the polymerase used can read across uracil (see Subheading 3.2.7 for details).
20. Consumables including 0.2-mL PCR tubes (single, 8 strip, 12 strip, 48 well or 96 well plate depending on size of experiment), 1.5-mL microcentrifuge tubes, 50-mL Falcon tubes.

2.2. Materials Needed Only for Protocol 1

1. 1× TE buffer: 10 mM Tris–HCl, 1 mM EDTA, pH 8.0. To prepare 50 mL of 1× TE, add 100 μL 0.5 M EDTA, pH 8.0 and 500 μL 1 M Tris, pH 8.0 to a 50-mL falcon tube and add ultrapure water to 50 mL.
2. 2× BindWash buffer (2× BWT): 2 M NaCl, 10 mM Tris–Cl, 1 mM EDTA, 0.05 % Tween-20, pH 8.0. Fill 20 mL of 5 M NaCl into a falcon tube and add 30 mL of 1× TE. Add 25 μL Tween-20 (see Note 1).

3. 1× BindWash buffer (1× BWT): 1 M NaCl, 10 mM Tris–Cl, 1 mM EDTA, 0.05% Tween-20, pH 8.0. Fill 10 mL of 5 M NaCl into a falcon tube and add 40 mL of 1× TE. Add 25 μL Tween-20 (see Note 1).
4. Invitrogen M-270 Dynabeads or MyOne C1 beads (Streptavidin beads).

2.3. Materials Needed Only for Protocol 2

1. (Optional) EB buffer (supplied with Qiagen MinElute PCR Purification kit): 10 mM Tris–HCl, pH 8.5.

3. Methods

3.1. Prepare Indexing Adapters

Set up the following hybridization reactions in 0.2-mL PCR tubes.

1. Prepare a master mix including 10 μL Adapter_1 (500 μM), 10 μL Adapter_1_rev (500 μM), 10 μL 10× Oligo hybridization buffer and 70 μL of ultrapure water.
2. Repeat step 1 with Adapter_2 and Adapter_2_rev in a separate 0.2-mL PCR tube.
3. Incubate the reactions in a thermal cycler using the following profile: 95°C 10 s, ramp to 14°C (0.1°/s) (see Note 2). This will produce double-stranded adapters (dsAdapter_1 and dsAdapter_2) at a concentration of 50 μM.

3.2. Protocol 1

Production of barcoded sequencing libraries from highly degraded, low copy number DNA extracts.

3.2.1. Blunt-End Repair

1. Prepare a master mix for the required number of reactions. The specific details of the composition of the reaction are provided in Table 3. Mix carefully by pipetting up and down or flicking the tube with a finger. Do not vortex after adding enzymes. Keep the master mix on ice if not immediately used to maintain full enzyme activity.
2. Add 20 μL master mix to 20 μL sample dissolved in EB, TE, or water to obtain a total reaction volume of 40 μL and mix. Incubate in a thermocycler for 15 min at 25°C followed by 5 min at 12°C (see Note 3).
3. Immediately purify the reaction over a Qiagen MinElute silica spin column according to the manufacturer's instructions. Elute in 20 μL 0.1× TE + 0.05 % Tween-20.

3.2.2. Adapter Ligation

1. Prepare a ligation master mix for the required number of reactions. Specific details of the composition of the reaction are provided in Table 4. Since PEG is highly viscous, vortex the

Table 3
Master mix for blunt-end repair (Protocol 1)

Reagent	Volume (μL) per sample	Final concentration in reaction
Water (add to 20 μL)	7.6	
Buffer Tango (10×)	4	1×
dNTPs (2.5 mM each)	1.6	100 μM each
ATP (10 mM)	4	1 mM
T4 Polynucleotide kinase (10 U/μL)	2	0.5 U/μL
T4 Polymerase (5 U/μL)	0.8	0.1 U / μL

Table 4
Master mix for adapter ligation (Protocol 1)

Reagent	Volume (μL) per sample	Final concentration in reaction
Water (add to 19 μL)	10	
T4 Ligase buffer (10×)	4	1×
PEG-4000 (50%)	4	5%
T4 Ligase (5 U/μL)	1	0.125 U/μL

master mix before adding T4 ligase and mix gently thereafter. White precipitate may be present in the ligation buffer after thawing. Heat the buffer vial briefly to 37°C and vortex until the precipitate has dissolved.

2. Combine the eluate from step 3 with 0.5 μL of dsAdapter_1 (50 μM) and 0.5 μL of dsAdapter_2 (50 μM). Mix thoroughly and spin down. Add 19 μL of master mix to obtain a total reaction volume of 40 μL and mix. Visually verify that all reaction components are mixed well. Incubate for 1 h at 22°C in a thermal cycler (see Note 4).
3. Purify the reaction over a Qiagen MinElute silica spin column according to the manufacturer's instructions, but perform two PE washing steps. Elute in 25 μL 1× TE (without Tween-20). Incubate the elution buffer on the silica membrane for 5 min before spinning it through (see Note 5).

3.2.3. Prepare Beads

1. Resuspend stock solution of MyOne C1 streptavidin beads by gentle mixing. Use 15 μL bead suspension (per sample). Add 100 μL 2× BWT buffer, mix by pipetting and place on magnetic rack for 1 min. After discarding the supernatant, repeat the previous step and elute the beads in 25 μL 2× BWT buffer.

3.2.4. Library Immobilization

1. Add 25 μL eluate from step 6 to 25 μL bead suspension (from step 7). Incubate 15 min at room temperature and occasionally shake the tube/plate.
2. Pellet the supernatant using the magnetic rack and discard the supernatant.
3. Wash 4 times with 100 μL 1× BW buffer to fully remove unincorporated dsAdapter_1 and dsAdapter_1 dimers. Change tube/plate after wash 1 and 3. Remove the supernatant from the last washing step shortly before adding the fill-in master mix (see Note 6).

3.2.5. Adapter Fill-In

1. Prepare a fill-in master mix for the required number of samples. Specific details of the composition of the reaction are provided in Table 5.
2. Resuspend the bead pellet in 50 μL fill-in master mix. Incubate at 37°C for 20 min in a thermal cycler.
3. Pellet the beads using the magnetic rack and discard supernatant.
4. Wash the beads twice with 100 μL 1× BWT buffer. Completely remove the supernatant!

3.2.6. Library Elution

1. Add 20 μL 0.1× TE (without Tween 20) and boil at 95°C for 3 min.
2. Place on magnetic plate for 1 min and transfer supernatant into a fresh, preferably siliconized tube (see Note 7). The eluate is the single stranded, partly (Illumina Solexa) or fully barcoded (Roche 454) library with truncated adapters.

Table 5
Adapter fill-in master mix (Protocol 1)

Reagent	Volume (μL) per sample	Final concentration in reaction
Water (add to 50 μL)	38	
Thermopol buffer (10×)	5	1×
dNTPs (2.5 mM each)	5	250 μM each
Bst Polymerase (8 U/μL)	2	16 U

3.2.7. Library Quantification

1. Perform quantitative PCR (qPCR) using 1 μL of truncated library and quant_primer_F and quant_primer_R. This is to get an estimate of your copy number before amplification with the extension primers to avoid over-amplification of the libraries. Check qPCR amplicon on gel to identify potential adapter dimers. For a detailed protocol see (11). As ancient DNA molecules often contain uracil as a result of postmortem cytosin deamination, it is essential for accurate quantification to use a polymerase that can read across uracil, such as Amplitaq Gold (Applied Biosystems). For a list of suitable polymerases see (12).

3.2.8. Extension and Amplification of Libraries

1. Determine the number of PCR cycles to saturation from the amplification plots of the qPCR.
2. Prepare an amplification master mix. Specific details of the composition of the reaction are provided in Table 6.
3. Amplify the libraries in a thermal cycler to just below saturation to avoid the formation of heteroduplexes and other PCR artifacts that may interfere with downstream applications. The thermal profile is as follows:

95°C for 12:00 min	
94°C for 0:30 min 58°C for 0:30 min 72°C for 1:00 min	Repeat for *N* cycles as determined by qPCR
72°C for >10:00 min	
10°C for hold	

Table 6
Master mix for extension and amplification of libraries (Protocol 1)

Reagent	Volume (μL) per sample	Final concentration in reaction
Water (add to 50 μL)	18	
10× buffer (10×)	5	1×
25 mM $MgCL_2$	5	2.5 mM
dNTPs (25 mM each)	0.5	0.25 mM each
ext_ primer_F (10 μM)	1	0.2 μM
ext_ primer_R (10 μM)	1	0.2 μM
Amplitaq Gold	0.5	2.5 U
Template	19	

4. Load 4 μL of the amplified PCR products on a gel to check for potential adapter dimers. Then purify the reaction using the Agencourt AMPure XP DNA purification kit according to the manufacturer's instructions (also see (3)). Elute and store DNA in 20 μL 0.1× TE.

3.3. Protocol 2

The second protocol is designed for barcoding preamplified multiplex PCR products, but can also be used for barcoding regular PCR products. It should be noted that the protocol was initially designed for and tested on Roche's 454 platform (13), but should theoretically be compatible with Illumina's solexa platform by interchanging the respective adapters and primers. Since the protocol uses preamplified (or even fully amplified) PCR product as template material, smaller volumes can be used than in protocol 1. Therefore, all reaction volumes have been cut down to 30 μL reactions to be more cost efficient.

3.3.1. Blunt-End Repair

1. Prepare a master mix for the required number of reactions. Specific details of the composition of the reaction are provided in Table 7. Mix carefully by pipetting up and down or flicking the tube with a finger. Do not vortex after adding enzymes. Keep the master mix on ice if not immediately used to maintain full enzyme activity.
2. Add 15 μL master mix to 15 μL sample (dissolved in EB, TE, or water; see Note 3) to obtain a total reaction volume of 30 μL and mix. Incubate in a thermal cycler for 15 min at 25°C followed by 5 min at 12°C (see Note 3).
3. Immediately purify the reaction over a Qiagen MinElute silica spin column (for PCR products shorter than 100 bp) or the Agencourt AMPure XP DNA purification kit (for PCR products longer than 100 bp) according to the manufacturer's instructions. Elute in 15 μL 0.1× TE + 0.05% Tween-20.

Table 7
Master mix for blunt-end repair (Protocol 2)

Reagent	Volume (μL) per sample	Final concentration in reaction
Water (add to 15 μL)	5.7	
Buffer Tango (10×)	3	1×
dNTPs (2.5 mM each)	1.2	100 μM each
ATP (10 mM)	3	1 mM
T4 Polynucleotide Kinase (10 U/μL)	1.5	0.5 U/μL
T4 Polymerase (5 U/μL)	0.6	0.1 U/μL

3.3.2. Adapter Ligation

1. Prepare a ligation master mix for the required number of reactions. Specific details of the composition of the reaction are provided in Table 8. Since PEG is highly viscous, vortex the master mix before adding T4 ligase and mix gently thereafter. White precipitate may be present in the ligation buffer after thawing. Heat the buffer vial briefly to 37°C and vortex until the precipitate has dissolved.
2. Combine the eluate from step 3 with 2 μL dsAdapter_1 (50 μM) and 2 μL dsAdapter_2 (50 μM) (see Note 9). Mix thoroughly and spin down. Add 11 μL master mix to obtain a total reaction volume of 30 μL and mix. Visually verify that all reaction components are mixed well. Incubate for 1 h at 22°C in a thermal cycler (see Note 4).
3. Purify the reaction using the Agencourt AMPure XP DNA purification kit according to the manufacturer's instructions (also see (3)) and elute in 15 μL 0.1× TE+0.05% Tween-20.

3.3.3. Adapter Fill-In

1. Prepare a fill-in master mix for the required number of samples. Specific details of the composition of the reaction are provided in Table 9.

Table 8
Master mix for adapter ligation (Protocol 2)

Reagent	Volume (μL) per sample	Final concentration in reaction
Water (add to 11 μL)	4.25	
T4 Ligase buffer (10×)	3	1×
PEG-4000 (50%)	3	5 %
T4 Ligase (5 U/μL)	0.75	0.125 U / μL

Table 9
Adapter fill-in master mix (Protocol 2)

Reagent	Volume (μL) per sample	Final concentration in reaction
Water (add to 15 μL)	8	
Thermopol buffer (10×)	3	1×
dNTPs (2.5 mM each)	3	250 μM each
Bst Polymerase (8 U/μL)	1	8 U

2. Add 15 μL master mix to 15 μL eluate from step 6 to obtain a total reaction volume of 30 μL and mix. Incubate in a thermal cycler for 20 min at 37°C (use heated lid).
3. Immediately purify the reaction using the Agencourt AMPure XP DNA purification kit and elute in 25 μL 0.1× TE or EB without Tween-20.

3.3.4. Library Quantification (Optional Step)

1. Perform quantitative PCR (qPCR) using suitable qPCR reagents (e.g., the HS SYBR® Green qPCR Kit, New England Biolabs) following the manufacturer's instructions and adding 1 μL of truncated library and quant_primer_F1 and quant_primer_R1 (see Note 10). This is to get an estimate of your copy number before amplification with the extension primers to avoid over-amplification of the libraries. Run out the qPCR amplicon on a gel to identify potential adapter dimers. For a detailed protocol see (11). If primer dimers are visible on the gel, purify the reaction again using the Agencourt AMPure XP DNA purification kit.

3.3.5. Extension and Amplification of Libraries

If step 1 of Subheading 3.3.4 (above) was performed, determine the number of PCR cycles to saturation from the amplification plots of the qPCR. In the absence of quantification results, perform 15 PCR cycles.

1. Prepare a fill-in master mix for the required number of samples. Specific details of the composition of the PCR are provided in Table 10.
   Thermal profile:

95°C for 12:00 min	
94°C for 0:30 min 60°C for 0:30 min 72°C for 0:40 min	15 cycles or *N* cycles as determined from optional qPCR (see Subheading 3.3.4)
72°C for 10:00 min	
10°C for hold	

2. Purify the reactions using the Agencourt AMPure XP DNA purification kit. Elute and store the double-stranded libraries in 20 μL 0.1× TE + 0.05 % Tween-20.

3.3.6. Final Quantification of Libraries

1. Perform qPCR using, e.g., the HS SYBR® Green qPCR Kit (New England Biolabs) following the manufacturer's instructions and adding 1 μL of the double-stranded final library and Amp_primer_F and Amp_primer_R (see Note 11). Run out the qPCR amplicons on a gel to identify potential adapter dimers. For a detailed protocol see (11). If primer dimers are visible on the gel, purify the reaction again using the Agencourt

Table 10
Master mix for extension and amplification of libraries (Protocol 2)

Reagent	Volume (μL) per sample	Final concentration in reaction
Water (add 20 μL)	4.95	
10× buffer (10×)	2	1×
25 mM $MgCL_2$	1.6	2 mM
dNTPs (25 mM each)	0.2	0.25 mM each
ext_ primer_F (10 μM)	0.5	0.25 μM
ext_ primer_R (10 μM)	0.5	0.25 μM
Amplitaq Gold (5 U/μL)	0.25	1.25 U
Template	10	

AMPure XP DNA purification kit. Further evaluate the success of the barcoding procedure by verifying the expected size-shift from your template DNA to the increased size of the final library (sum of template DNA and the length of both adapters) (see Note 12).

2. Following quantification, the double-stranded libraries can be pooled in equimolar ratios and submitted to the standard 454 sequencing pipeline.

4. Notes

1. A low concentration of Tween 20 in the elution buffer can prevent DNA molecules from sticking to the tube walls. As a result, less DNA is lost when transferring DNA between reaction tubes. Tween-20 also improves handling of SPRI and Streptavidin beads.
2. Alternatively to this protocol, the 5′ and 3′ adapters can also be adjusted to 100 μM each and mixed to produce a ready-to-use adapter mix. However, if both adapters carry barcodes, premixing the adapters will limit the number of possible barcode combinations.
3. Prolonged incubation may cause recessed instead of blunt ends and reduce ligation efficiency.
4. It is essential to combine DNA and adapters first prior to adding the master mix containing the ligase. If the adapters are

added to the master mix containing the ligase, large amounts of adapter dimers may form. Adding the master mix to the DNA before adding the adapters may increase the number of chimeric, concatenated target molecules.

5. To reduce the amount of unincorporated adapters, two PE washes are recommended at this step.
6. As very low copy number aDNA extracts are difficult to quantify before adapter ligation, the exact amount of adapters needed for the ligation cannot easily be calculated. Therefore, an excess of adapters is usually used. However, excess adapters and adapter dimers can significantly reduce the number of reads on target (reads that map to the target regions) produced by the NGS instrument. It is therefore essential to reduce the amount of dimers and unincorporated adapters in the library as much as possible by performing the washing procedure described under Subheading 3.2.4.
7. Siliconized tubes prevent DNA from sticking to tube walls. Thus less DNA is lost through repeated freeze/thaw cycles.
8. In case of using fully amplified PCR products (>10 ng/μL) as a template, 15 μL of a 1:10 dilution of the purified PCR product will be sufficient.
9. To avoid an increased number of chimeric, concatenated target molecules due to self-ligation, the adapter concentration in the reaction may be increased up to fourfold, or the template concentration decreased (see Note 8).
10. Note that only primers "quant_prim_F1" and "quant_prim_R1" can be used, since the library is still truncated and does not yet contain the "quant_prim_F2" and "quant_prim_R2" priming sites.
11. At this stage, "quant_primer" and "Amp_primer" pairs can be used for quantification. For the 454 platform, "Amp_prim_F" and "Amp_prim_R" are identical in their sequence to the emPCR primers provided by Roche and should be preferred at this point, since they will best mimic the downstream emulsion PCR reaction and usually give more consistent and reproducible quantification results.
12. On the gel, the final libraries might appear to comprise DNA fragments that are longer in size than expected. This is most likely the result of heteroduplexes (library molecules that consist of two noncomplementary strands, originating from different template molecules which only hybridize in the flanking adapter regions) migrating slower than perfectly double-stranded molecules in the gel.

Acknowledgments

This work was supported by the Max Planck Society, the German Research Foundation (DFG), the Allan Wilson Centre for Molecular Ecology and Evolution, the University of Otago and Pennsylvania State University.

References

1. Sanger F, Nicklen S, Coulson AR (1977) DNA sequencing with chain-terminating inhibitors. Proc Natl Acad Sci U S A 74:5463–5467
2. Knapp M, Hofreiter M (2010) Next generation sequencing of ancient DNA: requirements, strategies and perspectives. Genes 1: 227–243
3. Meyer M, Stenzel U, Hofreiter M (2008) Parallel tagged sequencing on the 454 platform. Nat Protoc 3:267–278
4. Meyer M, Kircher M (2010) Illumina sequencing library preparation for highly multiplexed target capture and sequencing. Cold Spring Harb Protoc. doi:10.1101/pdb.prot5448
5. Margulies M, Egholm M, Altman WE et al (2005) Genome sequencing in microfabricated high-density picolitre reactors. Nature 437:376–380
6. Maricic T, Pääbo S (2009) Optimization of 454 sequencing library preparation from small amounts of DNA permits sequence determination of both DNA strands. Biotechniques 46:51–57
7. Briggs AW, Good JM, Green RE et al (2009) Targeted retrieval and analysis of five Neandertal mtDNA genomes. Science 325:318–321
8. Meyerhans A, Vartanian JP, Wain-Hobson S (1990) DNA recombination during PCR. Nucleic Acids Res 18:1687–1691
9. Nikiforov TT, Rendle RB, Kotewicz ML, Rogers YH (1994) The use of phosphorothioate primers and exonuclease hydrolysis for the preparation of single-stranded PCR products and their detection by solid-phase hybridization. PCR Methods Appl 3:285–291
10. Craig DW, Pearson JV, Szelinger S et al (2008) Identification of genetic variants using barcoded multiplexed sequencing. Nat Methods 5:887–893
11. Meyer M, Briggs AW, Maricic T et al (2008) From micrograms to picograms: quantitative PCR reduces the material demands of high-throughput sequencing. Nucleic Acids Res 36:e5
12. Heyn P, Stenzel U, Briggs AW et al (2010) Road blocks on paleogenomes—polymerase extension profiling reveals the frequency of blocking lesions in ancient DNA. Nucleic Acids Res 38:e161
13. Stiller M, Knapp M, Stenzel U et al (2009) Direct multiplex sequencing (DMPS)-a novel method for targeted high-throughput sequencing of ancient and highly degraded DNA. Genome Res 19:1843–1848

Chapter 20

Case Study: Targeted high-Throughput Sequencing of Mitochondrial Genomes from Extinct Cave Bears via Direct Multiplex PCR Sequencing (DMPS)*

Mathias Stiller

Abstract

Here I describe the use of a recently developed technique for targeted high-throughput sequencing of highly degraded DNA by direct multiplex PCR sequencing (DMPS) that was used to amplify 31 near-complete mitochondrial genomes of the extinct cave bear (*Ursus spelaeus*). DMPS couples multiplex PCR with the generation of barcoded sequencing libraries to be sequenced in parallel on a high-throughput sequencing platform. DMPS makes it possible to generate large amounts of targeted DNA sequence data simultaneously from multiple degraded samples such as fossil remains. In this chapter, I describe an experiment that uses DMPS with different primer sets and on both modern and ancient DNA templates.

Key words: Cave bear, *Ursus spelaeus*, Mitochondrial genome, Multiplex PCR, Roche 454 FLX platform, High-throughput sequencing, Target enrichment

1. Introduction

Using traditional PCR, cloning and Sanger sequencing, and an ancient specimen with average DNA preservation, many amplifications would be necessary to obtain the complete mitochondrial genome sequence from a single individual (1). The process would necessarily consume a large amount of irreplaceable tissue (e.g., bone) in order to provide sufficient amounts of DNA extract for all the amplification reactions. A two-step multiplex PCR approach, such as that described in Chapter 17 (13) dramatically reduces

*Note: In the case study presented in this chapter, I describe the amplification and sequencing of whole mitochondrial genomes using a combined approach of the methods presented in Chapters 15 (12), and 19 (11). I discuss specific challenges associated with using this method to amplify and sequence modern and ancient DNA templates. For more information, see the original publication of the scientific results in Stiller et al. (2009) (4).

Beth Shapiro and Michael Hofreiter (eds.), *Ancient DNA: Methods and Protocols*, Methods in Molecular Biology, vol. 840,
DOI 10.1007/978-1-61779-516-9_20, © Springer Science+Business Media, LLC 2012

both the amount of time and extract required to produce these data. Tagging protocols (2, 3) can further simplify the process, where all second-step multiplex PCR products are barcoded, pooled, and converted into a sequencing library, and these libraries sequenced collectively using a high-throughput sequencing platform. This process can be even further simplified by directly coupling the multiplex PCR and the barcoding and library preparation steps (4). Using this approach, called direct multiplex PCR sequencing (DMPS), barcoded sequencing adapters are immediately added to the first-step multiplex PCR reaction, and all of the second-step reactions can be omitted. DMPS enables long, continuous sequences to be obtained rapidly from multiple individuals.

I describe the use of DMPS to generate 31 near-complete mitochondrial genomes of cave bears (*U. spelaeus*). Until their extinction about 25,000 years ago, cave bears were one of the most abundant mammalian species in Europe and Asia (5). Analyses of their remains have revealed a large amount of morphological and genetic diversity that has been loosely divided into three major lineages: *U. spelaeus*, *U. ingressus*, and *U. deningeri kudarensis* (6–8). Because of the challenges associated with extracting and amplifying ancient DNA, phylogenies were based on only short (ca 285 base pair (bp)) fragments of the mitochondrial control region, and were not well resolved (6–8). The mitochondrial genomes generated using DMPS were used to resolve the phylogenetic relationships among the major cave bear lineages.

2. Materials and Methods

As part of a previously published project (4), ancient DNA (aDNA) was extracted from 110 cave bear bone or tooth specimens representing most of the species' geographical range using a silica-based extraction method (9). To evaluate the DNA preservation, PCR amplification of a 175 bp fragment of the mitochondrial control region was attempted from all extracts using the primers 2620F (5′-GCCCCATGCATATAAGCATG-3′) and 2558R (5′-GGAGCGAGAGGTACACGT-3′). Based on these results, we selected specimens that were sufficiently well preserved for further processing.

We used Multiplex PCR to amplify the whole mitochondrial genome of the well-preserved specimens using two, nonoverlapping sets of primers. Each of the two primer sets consisted of 64 primer pairs and targeted fragments of between 150 and 180 bp in length. We designed all primers using the online tool primer3 (http://frodo.wi.mit.edu/primer3/). Because of the large number of combinations of primers and possibilities for negative interaction between them, it would have been a significant challenge to create and follow a robust optimization strategy. Therefore, we took no particular care to avoid primer dimer formation between primers of different fragments in each set. We then used each primer set in standard multiplex PCRs containing 2 U AmpliTaq Gold DNA polymerase, 1× AmpliTaq Gold buffer, 2.5 mM $MgCl_2$, 250 μM of each dNTP, 0.8 mg/mL BSA, and 150 nM of each primer. The reactions were cycled with an activation step of 12 min at 95°C, followed by 20 cycles of denaturation at 94°C for 30 s, annealing at 53°C for 30 s, and elongation at 72°C for 30 s, with a final extension step at 72°C for 10 min.

Next, we converted one multiplex PCR for each primer set for each sample (two libraries per sample) into a barcoded sequencing library using the barcoding "protocol 2" for pre-amplified DNA as described in Chapter 19 ((11); Fig. 1). This approach directly couples the barcoding protocol and the library preparation process by including the barcode sequence in the adapter sequence. We then quantified all libraries using quantitative PCR (qPCR). According to the qPCR results, we pooled the libraries in equimolar ratios and sequenced them simultaneously on a small (1/16th) lane of a 454 FLX sequencing plate. After sequencing, we sorted all of the obtained reads according to their barcode sequence, in this case the first seven bases of the sequencing reads. In the ideal scenario,

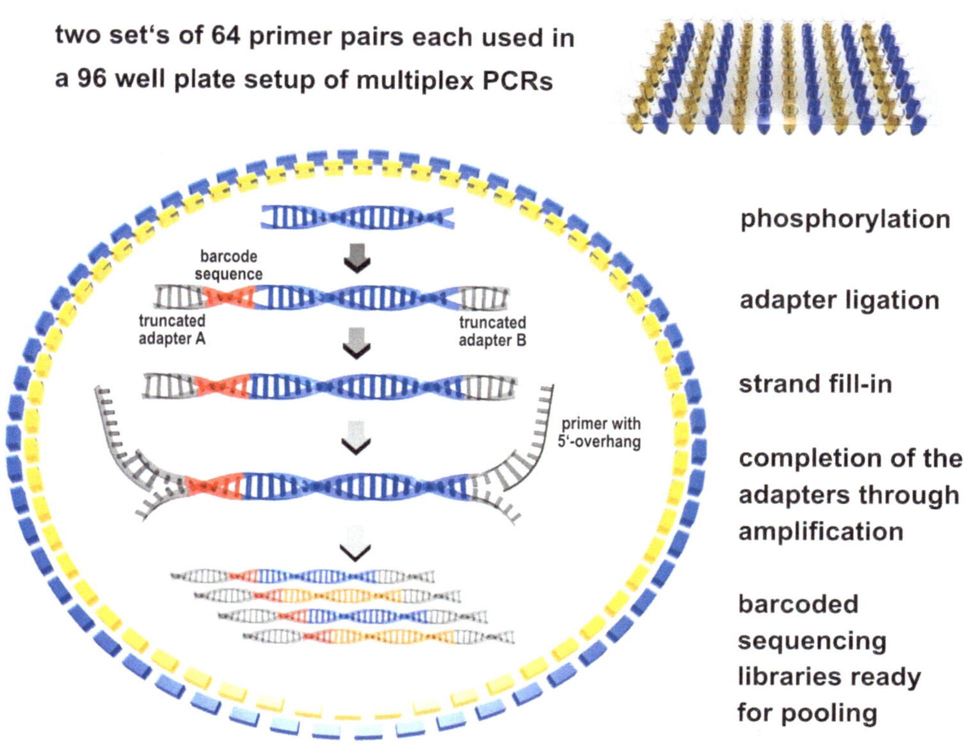

Fig. 1. Schematic overview of the combined protocol coupling first-step multiplex PCR presented in Chapter 17 directly to barcoding protocol 2 of Chapter 19.

all barcodes would be represented evenly in terms of number of reads. Errors introduced during upstream steps, such as incorrect quantification in qPCR, errors in the dilution steps of the pooling procedure, or simple pipetting errors, can, however, result in an under- or overrepresentation of barcodes in the final sequencing output.

After verifying a fairly balanced representation of barcodes in the library pool and a sufficient enrichment of the target fragments, we sequenced four multiplex PCRs (the "odd" and "even" sets in replicates, respectively) for each of the selected cave bear samples on a full 454 FLX run. We then performed a second round of multiplex PCR in order to fill remaining gaps in the cave bear mitochondrial genome sequences. In this multiplex PCR, however, the primer sets contained only those primer pairs that flanked missing sequence data. To compensate for the reduced number of targeted fragments and to ensure amplification of the target fragments above the environmental DNA background, we increased the number of cycles in the PCR from 20 to 25.

3. Results and Discussion

Fifty-six of the one hundred and ten cave bear specimens tested showed sufficiently well-preserved DNA to be used in multiplex PCR. After 20 cycles, the reactions were converted into barcoded sequencing libraries, quantified and pooled in equimolar ratios, and sequenced on a small (1/16th) 454 FLX lane. Analysis of these initial sequencing results revealed differences between the samples, either in DNA preservation or in the amount of contamination with exogenous DNA (e.g., fungal and bacterial DNA). The proportion of sequence reads that matched the target fragments varied widely among the 56 specimens used, from 1% to 100%.

As only 1% of reads matching target fragments is insufficient to compile a consensus sequence, we continued to process only those samples that were the best preserved. We applied an arbitrary cut-off in which we required at least 40% of the sequencing reads to have matched the targeted fragments in order to keep a sample in the experiment. Instead of applying this cut-off, one could have chosen to re-amplify the more poorly performing samples (those showing low amounts of endogenous DNA and/or high levels of contamination with exogenous DNA), this time increasing the number of PCR cycles to 25 or up to 30 cycles. Note that increasing the number of cycles will also increase the uneven representation among the target fragments in the reaction, due to differences in amplification efficiency among primer pairs. Too few cycles, however, may be insufficient to enrich for the target fragments over the environmental background DNA. It is therefore highly recommended to determine the ratio of reads matching target fragments to reads matching environmental background DNA prior to final deep sequencing.

In this case, we continued to process 31 of the 56 samples that met our preservation criterion. Based on the obtained output, 112 of the 128 target fragments were covered by sequencing reads on average among the 31 samples. Thus, based on only one full run of the 454 FLX instrument, on average 87% of the mitochondrial genome was obtained from 31 individuals, representing more than 7 kilobases (kb) of replicated, overlapping sequence from all of the 31 individuals. With only one more round of gap filling, on average 96% of the mitochondrial genomes were covered, translating into ~10 kb of overlapping sequence from all individuals.

Phylogenetic analyses of the consensus sequences revealed a stable topology with very high statistical support, indicating strong evidence for the reciprocal monophyly of the three cave bear lineages (4).

DMPS has also been used successfully in experiments to amplify whole mitochondrial genomes from a modern polar bear and a fossil mammoth, as well as to amplify multiple nuclear loci from a modern African elephant (4). In addition to using different primer sets designed for the respective species and target loci, the only other modification to the protocol described above was, when modern samples were used, to lower the number of PCR cycles from 20 to 15.

These results show that no extensive optimization of primer sets is necessary to successfully apply DMPS to ancient or modern DNA sequencing experiments. Further, DMPS, like traditional PCR, offers full single-molecule sensitivity (10), as no pretreatment of the aDNA extract (e.g. library preparation) is necessary prior to amplification. The protocol is therefore an easy-to-implement, robust, and cost-efficient way to quickly retrieve many kb of homologous sequence data from large numbers of highly degraded samples, such as fossil remains and poorly preserved samples from museum, forensic, and medical collections.

Acknowledgments

I thank M Meyer and M Hofreiter for help throughout the research project; B Hoeffner and A Aximu for running the 454 sequencer; G Baryshnikov, H Bocherens, A Grandal d'Anglade, B Hilpert, T Kutznetsova, S Münzel, R Pinhasi, G Rabeder, W Rosendahl, and E Trinkaus for providing samples; K Finstermeier for help with the figure and the Max Planck Society and National Science Foundation (award ANS-0909456) for financial support.

References

1. Bon C, Caudy N, de Dieuleveult M, Fosse P, Philippe M, Maksud F, Beraud-Colomb E, Bouzaid E, Kefi R, Laugier C, Rousseau B, Casane D, van der Plicht J, Elalouf JM (2008) Deciphering the complete mitochondrial genome and phylogeny of the extinct cave bear in the Paleolithic painted cave of Chauvet. Proc Natl Acad Sci U S A 105:17447–17452
2. Binladen J, Gilbert MT, Bollback JP, Panitz F, Bendixen C, Nielsen R, Willerslev E (2007) The use of coded PCR primers enables high-throughput sequencing of multiple homolog amplification products by 454 parallel sequencing. PLoS One 2:e197
3. Meyer M, Stenzel U, Hofreiter M (2008) Parallel tagged sequencing on the 454 platform. Nat Protoc 3:267–278
4. Stiller M, Knapp M, Stenzel U, Hofreiter M, Meyer M (2009) Direct multiplex sequencing (DMPS)—a novel method for targeted high-throughput sequencing of ancient and highly degraded DNA. Genome Res 19: 1843–1848
5. Pacher M, Stuart AJ (2009) Extinction chronology and palaeobiology of the cave bear (*Ursus spelaeus*). Boreas 38:189–206
6. Knapp M, Rohland N, Weinstock J, Baryshnikov G, Sher A, Nagel D, Rabeder G, Pinhasi R, Schmidt HA, Hofreiter M (2009) First DNA sequences from Asian cave bear fossils reveal deep divergences and complex phylogeographic patterns. Mol Ecol 18:1225–1238
7. Hofreiter M, Rabeder G, Jaenicke-Despres V, Withalm G, Nagel D, Paunovic M, Jambresic G, Pääbo S (2004) Evidence for reproductive isolation between cave bear populations. Curr Biol 14:40–43
8. Rabeder G, Hofreiter M, Withalm G (2004) The systematic position of the Cave Bear from Potocka zijalka (Slovenia). Mitt Komm Quartärforsch Österr Akad Wiss 13:197–200
9. Rohland N, Hofreiter M (2007) Ancient DNA extraction from bones and teeth. Nat Protoc 2:1756–1762
10. Dear PH, Cook PR (1993) Happy mapping: linkage mapping using a physical analogue of meiosis. Nucleic Acids Res 21:13–20
11. Knapp M, Stiller M, Meyer M (2011) Generating barcoded libraries for multiplex high-throughput sequencing. In: Shapiro B, Hofreiter M (eds) Ancient DNA. Springer, New York
12. Fulton TL, Stiller M (2011) PCR amplification, cloning and sequencing of ancient DNA. In: Shapiro B, Hofreiter M (eds) Ancient DNA. Springer, New York
13. Stiller M, Fulton TL (2011) Multiplex PCR amplification of ancient DNA. In: Shapiro B, Hofreiter M (eds) Ancient DNA. Springer, New York

Chapter 21

Target Enrichment via DNA Hybridization Capture

Susanne Horn

Abstract

Recent advances in high-throughput DNA sequencing technologies have allowed entire nuclear genomes to be shotgun sequenced from ancient DNA (aDNA) extracts. Nonetheless, targeted analyses of specific genomic loci will remain an important tool for future aDNA studies. DNA capture via hybridization allows the efficient exploitation of current high-throughput sequencing for population genetic analyses using aDNA samples. Specifically, hybridization capture allows larger data sets to be generated for multiple target loci as well as for multiple samples in parallel. "Bait" molecules are used to select target regions from DNA libraries for sequencing. Here we present a brief overview of the currently available hybridization capture protocols using either an in-solution or a solid-phase (immobilized) approach. While it is possible to purchase ready-made kits for this purpose, I present a protocol that allows users to generate their own custom bait to be used for hybridization capture.

Key words: Ancient DNA, Target enrichment, Hybridization, DNA capture, Bait, High-throughput sequencing

1. Introduction

Shotgun sequencing using next-generation sequencing techniques has been used to sequence entire genomes of ancient specimens (1–3). However, this approach remains prohibitively expensive for many users, and generally provides data from only a single specimen. Analyses of ancient populations generally do not focus on complete genome sequences, but instead on selected genomic loci that can be targeted from many individuals.

In many ancient DNA (aDNA) extracts, DNA fragments representing the target loci are present at very low copy-number compared to sequences of contaminating exogenous DNA. Such experiments therefore require an enrichment step, where the amount of target DNA is increased in a library to be sequenced, relative to nontarget DNA. Enrichment is most often achieved via

Beth Shapiro and Michael Hofreiter (eds.), *Ancient DNA: Methods and Protocols*, Methods in Molecular Biology, vol. 840, DOI 10.1007/978-1-61779-516-9_21, © Springer Science+Business Media, LLC 2012

polymerase chain reaction (PCR). This approach, however, is currently being superseded by enrichment strategies that capture DNA by hybridization (4–7). In hybridization capture approaches, a genomic library is first prepared from an aDNA extract and DNA bait molecules representing the target sequence are added to the library. The target DNA molecules in the library will hybridize with the added bait molecules and can then be pulled down out of the library for sequencing. DNA hybridization capture has several advantages compared to traditional PCR. First, while mismatches can prevent the binding of primers in PCR, mismatches are less detrimental for hybridization, making hybridization a useful method to enrich for DNA where the sequence of the ancient specimen is not exactly known. This can also be important when molecules with damage-induced base modifications may inhibit primer binding (8, 9). Second, hybridization is less sensitive to contamination than traditional PCR. While PCR selects for full-length amplicons and therefore tends to amplify longer molecules preferentially (which may be modern DNA contaminants), hybridization targets all lengths of starting molecules more equally. Third, nuclear mitochondrial insertions (numts) may be amplified preferentially by PCR if the primer binding conditions allow. Hybridization, however, should preferentially enrich for the most common fragment, which will be the much higher copy-number mitochondrial sequence. One potential drawback of hybridization capture is the loss of target molecules during library preparation. This is not a problem for PCR, which is theoretically able to begin the amplification process from a single starting molecule. Therefore, it is highly recommended that not all of the aDNA extract is used in a single enrichment experiment, but that some is saved for replication if necessary.

The choice of sequencing platform will determine what type of library will need to be prepared prior to enrichment (see Table 1). This choice may depend on the size of the sequence fragment to be targeted and the number of samples to be processed. Hybridization capture can be used to enrich for fragments ranging in length from a few hundred bases to many megabases (Mb) in size. When the sequencing is complete, only a fraction of the sequencing reads will map to the desired target region, and this also needs to be considered when planning the amount of sequence data that will be required. In previous work, enrichment rates for aDNA varied considerably across experiments: between 18 and 40% of reads could be mapped to a target region of a Neandertal mitochondrial genome (10); 37% of reads mapped to targeted nuclear regions of Neandertals (7); and around 20% of reads mapped to a targeted 500-base-pair (bp) region of the mitochondrial control region of beavers (*Castor fiber*) (see Chapter 22). Accordingly, deeper sequencing of these libraries may be necessary to reach sufficient coverage (e.g., 20× coverage might be the desired coverage of a resequencing experiment) for a targeted region.

Table 1
The sequencing throughput required and hybridization capture protocol recommended will depend on the desired target region size. Additionally, the expected percentage of reads that map to target and the desired coverage will influence the amount of data (in Megabases, Mb, or kilobases, kb) that need to be sequenced. The amount of endogenous DNA is difficult to estimate for aDNA extracts and influences the percentage of reads that will map to target. Therefore, the values presented here should be seen as minimums

Region size	2.5 Mb (e.g., any nuclear region)	16 kb (e.g., mitochondrial genome)	500 bp (e.g., mitochondrial control region)
Percentage of reads that map to target	30	30	20
Desired coverage	20×	20×	20×
Bp to sequence	167 Mb	1 Mb	50 kb
Recommended hybridization capture protocol	NimbleGen arrays, Agilent Sure Select, MYselect custom probes	Custom self-made bait prepared using long-range PCR and hybridization in solution	Custom self-made bait prepared using PCR and hybridization in solution; Primer extension capture (PEC)

Blocking oligonucleotides may also be used as part of the hybridization protocol. Because all of the DNA fragments in the library will have the same universal adapters ligated to their ends, they may hybridize to each other during the enrichment protocol, forming long nonsensical chains. Blocking oligonucleotides cover the ends of the sequences that contain the adapters, preventing accidental hybridization between adapters and thereby also preventing nontarget DNA molecules from being pulled down along with target molecules. The particular blocking oligonucleotides required will depend on the sequencing platform to be used.

An outline of different enrichment protocols is provided in Table 2. The hybridization mixture (containing the DNA library and the bait) can either be incubated in solution or immobilized in a solid phase on arrays or beads. Hybridization in solution and immobilization on beads both require conventional tubes and hybridization ovens, but incubation may also take place in a thermal cycler. If arrays are used, these need to be placed in special racks for rotation. Hybridization in solution requires a subsequent bead capture step, which is not necessarily required for either immobilization approaches. It has been suggested that hybridization in solution may be more efficient for libraries with fragment lengths shorter than 500 bp, as is expected for most aDNA libraries (5).

Table 2
Approaches of DNA hybridization capture for target enrichment prior to sequencing

Mode of hybridization	Hybridization in solution			Immobilized hybridization		
Immobilization	Capture on beads after hybridization			On Arrays		On beads
Approach no.	1	2	3	4	5	6
Technology	MYselect or Agilent SureSelect (5)	Self-made biotinylated oligos (15)	Primer extension capture (PEC) (10)	Agilent SureSelect DNA Capture Array	Roche Nimblegen sequence capture array (6)	Self-made biotinylated oligos (14)
Principle of enrichment	RNA bait is used to hybridize with the target and captured via attached biotin.	DNA bait is used to hybridize with the target and captured via attached biotin.	Primers hybridize, are elongated by a polymerase and captured via attached biotin.	DNA bait is bound to the microarray and hybridizes with the target.	DNA bait is bound to the microarray and hybridizes with the target.	DNA bait is bound to magnetic beads and hybridizes with the target.
Bait, or probes	RNA oligos	DNA stretches	DNA oligos	DNA oligos	DNA oligos	DNA stretches
Length of bait	120–200 bp	Up to thousands of bp	Around 30 bp	25–60 bp	Around 60 bp	Up to thousands of bp

Immobilization of the bait and tight physical clustering of bait molecules, as is common on arrays, may result in steric interference between target and nontarget molecules. The resulting pulldown of nontarget DNA could cause fewer sequencing reads to map to the desired target region. Finally, generating bait rather than purchasing bait may reduce the cost of the enrichment considerably. The protocol presented in this chapter describes how to generate self-made bait for longer and shorter contiguous targets. For later immobilization on streptavidin-coated beads, biotin needs to be introduced into the bait. This is achieved by the ligation of biotinylated adapters to sheared long-range PCR products. Alternatively, biotin can be introduced into shorter amplicons during a biotinylating PCR step.

Serial enrichments, where enrichment is performed more than once for a single library, can be applied to aDNA libraries that contain very low amounts of endogenous DNA compared to contaminating or background DNA (11). In such instances, a single enrichment step may be insufficient to provide adequate coverage of the target loci. The case study reported in Chapter 22 used two enrichment steps to target an approximately 500-bp stretch of mitochondrial DNA of beavers.

PCR amplification of libraries prior to, during, or after hybridization is all potentially problematic, but may be nonetheless unavoidable in many aDNA applications. Potential drawbacks of amplification include a skewed representation of the original library due to preferential amplification of certain molecules, jumping PCR artifacts such as chimeras, and additional errors introduced by polymerases. Commercial hybridization kits both in solution and on arrays require 3–4 μg of DNA in a library (12), which cannot be achieved from most ancient samples without amplification. In addition, amplification is generally necessary to produce sufficient DNA for sequencing on either the Illumina or SOLiD platforms (however, as little as 10^6 molecules per 1/16th lane may be sufficient for 454 sequencing). Thus, when using self-made bait or PEC primers (10) for hybridization (approaches no. 2, 3, and 6 in Table 2) with subsequent 454 sequencing, amplification steps may be avoidable. Because the potential problems associated with amplification are most likely to occur when the library quality is poor, care should be taken to select samples with the best quality and quantity of endogenous DNA, as may be identified using quantitative PCR.

Manufacturers provide a variety of arrays with capture probes made from DNA and RNA as well as in-solution capture kits (13) (see Note 1 and Table 2). Instead of purchasing a kit, however, DNA hybridization capture can be performed with standard tools in any molecular biology lab. I present a protocol to generate self-made bait to target genomic regions of interest. For this, it is possible to use the products of long-range PCR, thereby covering larger target regions. Bait can also be produced during a regular PCR for shorter targets. I provide a protocol for hybridizing a library to this self-made bait, and for the subsequent bead capture step that immobilizes the reaction. After elution, the enriched library can be directly sequenced on a high-throughput sequencing platform.

2. Materials

All reagents and plasticware should be sterile, DNA and DNAse free.

1. aDNA library.
2. If bait is to be produced from long-range PCR products:
   (a) Sonicator (e.g., Diagenode or Covaris).
   (b) Two complementary oligonucleotides, one of them 5′-biotinylated.
   (c) Oligo hybridization buffer (10×): 500 mM NaCl, 10 mM Tris-Cl (pH 8.0), 1 mM EDTA (pH 8.0).
   (d) Tango buffer (10×, e.g. Fermentas).
   (e) T4 DNA ligase (5 U/μL, supplied with 10× T4 DNA ligase buffer and 50% PEG-4000 solution).
   (f) T4 DNA polymerase (5 U/μL).
   (g) T4 polynucleotide kinase (10 U/μL).
   (h) ATP (100 mM).
   (i) *Bst* DNA polymerase, large fragment (supplied with 10× buffer).

or

2. If bait is to be produced by biotinylating PCR:
   (a) Biotin-dUTP (100 μM).
   (b) dNTP mix containing 25 mM of dATP, dGTP, and dCTP, but 24.5 mM of dTTP (mix 10 μL each of 100 mM dATP, dGTP, and dCTP with 9.8 μL 100 mM dTTP and 0.2 μL H_2O).
   (c) A polymerase incorporating biotinylated nucleotides (e.g., taq polymerase).
3. Streptavidin covered magnetic beads (e.g., Dynabeads M270, Invitrogen).
4. Tween-20.
5. EBT and TET: elution buffer from any kit and 1× TE buffer, including 0.05% Tween-20.

6. 1× bind and wash (BWT) buffer : 1 M NaCl, 10 mM Tris-Cl, 1 mM EDTA, 0.05% Tween-20, pH 8.0.
7. Hot wash (HW) buffer : 200 μL 10× PCR buffer, 200 μL $MgCl_2$ (25 mM), 1.6 mL H_2O.
8. Phusion High Fidelity PCR master mix (New England Biolabs).
9. SPRI beads (Agencourt AMPure XP) or MinElute kit (Qiagen).
10. Hybridization buffer and blocking agent (e.g., from the Agilent aCGH kit, Cat. no. 5188-5220).
11. Barrier/filter tips and PCR reaction tubes/plates.
12. DNA Spectrophotometer.
13. Magnetic rack for SPRI bead cleanups and capture with magnetic beads.
14. Hybridization oven.
15. A thermal cycler with heated lid.
16. Laboratory film (e.g., Parafilm).

3. Methods

Design and order blocking oligonucleotides. The sequences of blocking oligonucleotides correspond to the sequences of the respective adapters and can include ambiguity codes for barcodes which may vary between samples. Be sure to include the oligonucleotides to cover adapters in 5'–3' as well as in 3'–5' direction; examples are given in (11) (see Notes 2 and 3).

Produce biotinylated bait DNA for the enrichment using PCR products (14, 15). It is recommended to exclude repetitive regions from the PCR by placing the primers appropriately (see Note 4).

Generate DNA bait from long-range PCR products for target regions of kilobases in size, following manufacturers' instructions for long-range PCR

1. Prepare a biotinylated double-stranded adapter from two complementary oligonucleotides, one carrying a biotin at the 5′-end (see Note 5) in the following mix:
   (a) 40 μL of oligo 1 (500 μM).
   (b) 40 μL of oligo 2 (500 μM).
   (c) 10 μL 10× oligo hybridization buffer.
   (d) 10 μL H_2O.

   Heat the mixture to 95°C for 5 s, then ramp to 15°C at the rate of –0.1°C/sec. The resulting adapter has a concentration of 200 μM.

2. Fragment long-range PCR products with ultrasound in a sonicator twice for 7 minutes at "high" to obtain a fragment size of around 200–500 bp. Check the size of the obtained DNA on an agarose gel (1–2%) and, if necessary, repeat the fragmentation.
3. Purify the fragmented long-range PCR product using a MinElute column (see Note 6) and measure the concentration on a DNA spectrophotometer. Use up to 500 ng per reaction in the next step. Several reactions may be necessary to produce large amounts of bait exceeding 1 μg.
4. Set up a blunt-end repair, include per reaction:
   (a) 7.12 μL H_2O.
   (b) 7 μL Buffer Tango (10×).
   (c) 0.28 μL dNTPs (25 mM each).
   (d) 0.7 μL ATP (100 mM).
   (e) 3.5 μL T4 polynucleotide kinase (10 U/μL).
   (f) 1.4 μL T4 DNA polymerase (5 U/μL).

   Add 20 μL of master mix to 50 μL of purified long-range PCR product. Mix gently and incubate in a thermal cycler for 15 min at 25°C followed by 5 min at 12°C.
5. Place the reaction on ice or immediately perform a cleanup using a MinElute column and elute in 20 μL EBT.
6. Set up a master mix for the ligation of the biotinylated adapter, per reaction include:
   (a) 10 μL H_2O.
   (b) 4 μL T4 DNA ligase buffer (10×) (see Note 7).
   (c) 4 μL PEG-4000 (50%).
   (d) 1 μL adapter (200 μM).
   (e) 1 μL T4 DNA ligase (5 U/μL).

   Vortex the master mix before adding T4 DNA ligase and mix gently. Add 20 μL of master mix to each eluate from step 5 to obtain reaction volumes of 40 μL. Mix and incubate for 30 min at 22°C in a thermal cycler.
7. Purify the reaction using a MinElute column. Elute in 20 μL EBT.
8. Set up a master mix for the *Bst* fill-in, include per reaction:
   (a) 14.1 μL H_2O.
   (b) 4 μL ThermoPol reaction buffer (10×).
   (c) 0.4 μL dNTPs (25 mM each).
   (d) 1.5 μL *Bst* polymerase (8 U/μL).

   Add 20 μL of master mix to each eluate from step 7 to obtain reaction volumes of 40 μL. Mix and incubate in a thermal cycler for 20 min at 37°C.

9. MinElute purify the reaction and elute in 20 μL of EBT. Measure the concentration of DNA on a spectrophotometer. The bait DNA can be stored at −20°C for several months.

Biotin can also be introduced during the polymerization of target regions up to 1 kb in size during regular PCR. For each reaction set up a master mix with:

1. 3.2 μL $MgCl_2$ (25 mM).
2. 2 μL 10× PCR buffer.
3. 1 μL biotin-dUTP (100 μM) (see Note 8).
4. 0.2 μL dNTP mix.
5. 0.1 μL taq polymerase (5 U/μL).
6. 3 μL primer mix (10 μM each).
7. 9.5 μL H_2O.
8. 1 μL template DNA.

After 5–12 min of denaturation at 95°C, run the PCR for 35 cycles of 30 sec 94°C, 30 sec at the respective annealing temperature, and 1 min at 72°C for elongation.

After a MinElute cleanup, measure the bait solution on a spectrophotometer and store it at −20°C.

3.1. Hybridization

1. Prepare between 100 ng and 1 μg of library (to have a 10 times excess compared to the bait) for each hybridization reaction in 200 μL tubes or wells of a 96-well plate (see Note 9).
2. Set up a hybridization master mix with the following final concentrations:
   (a) 1× hybridization buffer.
   (b) 1× blocking agent.
   (c) 10–100 ng of bait (to achieve a final ratio bait:library of 1:10).
   (d) Blocking oligos (each 2 μM).
   (e) Water to 50 μL per hybridization reaction (accounting for library above).
3. Mix the master mix and add it to the library, resulting in 50 μL hybridization reactions (see Note 10).
4. After denaturation of the mixture at 95°C for 5 min, carry out the hybridization rotating at 65°C in a conventional hybridization oven (e.g., from SciGene) or in a thermal cycler (see Note 9). In the latter case, heat to 95°C for 5 min and then cool down to 65°C at 0.1°C/s.
5. Incubate at 65°C for 24 h or up to 48 h (see Note 11).

3.2. Immobilization of Target-Enriched Library

1. After hybridization, incubate the mixture with 5 μL magnetic streptavidin-coated beads for 20 min at room temperature (see Note 12).
2. Place the mixture into a magnetic rack to separate the magnetic beads from the supernatant (see Note 13).
3. Discard the supernatant, which contains the nontarget molecules.
4. Wash the beads 5 times using 1×BWT buffer, once in pre-warmed HW buffer at 50°C for 2 min, and once with 1×BWT.
5. Transfer the beads into a new tube and wash with 100 μL of TET.
6. Separate hybridized target molecules from the bait in 30 μL TE by incubation at 95°C for 5 min in a thermal cycler. The eluate containing the sequencing library enriched for target DNA is ready for amplification, quantification, and sequencing.

3.3. Serial Hybridization Captures

1. After the first hybridization capture, amplify the resulting library using the Phusion PCR master mix.
2. Purify the resulting amplicon with a MinElute column and use it in another round of hybridization capture starting at Subheading 3.1, step 1.

3.4. Amplification, Quantification, and Pooling Before Sequencing

1. Amplify the resulting library using the Phusion PCR master mix (see Note 14).
2. Quantify the enriched sequencing library with a spectrophotometer or by quantitative PCR.
3. Pool libraries of different samples (and negative controls, if applicable) in equimolar amounts for sequencing.

4. Notes

1. For hybridization on NimbleGen and Agilent arrays or when using the SureSelect in-solution kit, follow the protocols provided by the respective manufacturers (16, 17).
2. The use of blocking oligonucleotides is not mandatory but may increase the percentage of sequencing reads that map to the desired target region.
3. Since blocking oligonucleotides will have a length of more than 30 bp, most companies only provide them HPLC purified. This prevents shorter oligonucleotides (aborted synthesis products not reaching the full length) from being delivered along with the order. Special handling is not necessary.

Because the blocking oligonucleotides are combined in the hybridization mixture, potential cross-contamination is not a problem.

4. Repetitive regions should be excluded from your bait. Those could capture large amounts of repeats present in the library and swamp your sequencing results.
5. The two complementary oligonucleotides for the generation of a biotinylated adapter can be of arbitrary sequence; an example is given in (14).
6. SPRI beads (Agencourt Ampure XP kit) can be used for cleanup instead of MinElute columns if many samples have to be processed in parallel. See (18) for a detailed description of SPRI bead usage.
7. If white precipitate has formed in the 10× DNA ligase buffer after thawing, warm the buffer to 37°C and vortex until the precipitate has dissolved.
8. Increasing the amount of biotin-dUTP in relation to dTTP in the biotinylating PCR may yield a higher number of functional bait molecules. Up to 50% of dTTP can be substituted with its modified counterpart (19). This, however, will increase the cost of the experiment.
9. Enrichment in solution can be carried out for many samples in parallel in 96-well plates. These can be placed in a thermal cycler and should be incubated without rotation to minimize the chance of contamination between wells when improperly sealed.
10. When carrying out in-solution enrichment in tubes, be sure to seal the tubes properly and tape them with laboratory film (e.g., Parafilm) for the 65°C hybridization step under rotation.
11. In my experience, the throughput of DNA hybridization capture can be increased by using 96-well plates instead of single tubes and by shortening the hybridization time to 24 h, although the consequences of shortening the hybridization time have not been evaluated fully (17).
12. Dynabeads should not dry out; therefore remove buffers only immediately prior to the next pipetting step.
13. The addition of Tween (0.05%) to TE buffer facilitates the handling of streptavidin-coated magnetic beads. The beads will assemble closer to the magnet of the rack and will stick less to pipet tips and tube walls.
14. In case the enriched library will be sequenced on the 454 platform, a lower total amount of library is required for sequencing compared to Illumina and SOLiD. Thus, proceed with quantification of the library. Depending on the quantification results, the amplification step might not be necessary.

Acknowledgments

I would like to thank the Volkswagen foundation and the Max Planck society for funding and M Stiller for helpful comments on this manuscript.

References

1. Rasmussen M et al (2010) Ancient human genome sequence of an extinct Palaeo-Eskimo. Nature 463(7282):757–762
2. Green RE et al (2010) A draft sequence of the Neandertal genome. Science 328(5979): 710–722
3. Reich D et al (2010) Genetic history of an archaic hominin group from Denisova Cave in Siberia. Nature 468(7327):1053–1060
4. John JS, Quinn TW (2008) Rapid capture of DNA targets. Biotechniques 44(2): 259–264
5. Gnirke A et al (2009) Solution hybrid selection with ultra-long oligonucleotides for massively parallel targeted sequencing. Nat Biotechnol 27(2):182–189
6. Hodges E et al (2007) Genome-wide in situ exon capture for selective resequencing. Nat Genet 39(12):1522–1527
7. Burbano HA et al (2010) Targeted investigation of the Neandertal genome by array-based sequence capture. Science 328(5979): 723–725
8. Stiller M et al (2006) Inaugural article: patterns of nucleotide misincorporations during enzymatic amplification and direct large-scale sequencing of ancient DNA. Proc Natl Acad Sci U S A 103(37):13578–13584
9. Briggs AW et al (2007) Patterns of damage in genomic DNA sequences from a Neandertal. Proc Natl Acad Sci 104(37): 14616–14621
10. Briggs AW et al (2009) Targeted retrieval and analysis of five Neandertal mtDNA genomes. Science 325(5938):318–321
11. Meyer M, Kircher M (2010) Illumina sequencing library preparation for highly multiplexed target capture and sequencing. Cold Spring Harb Protoc 2010(6):pdb.prot5448. doi: 10.1101/pdb.prot5448
12. Teer JK et al (2010) Systematic comparison of three genomic enrichment methods for massively parallel DNA sequencing. Genome Res 20(10):1420–1431
13. Blow N (2009) Genomics: catch me if you can. Nat Methods 6(7):539–544
14. Maricic T, Whitten M, Pääbo S (2010) Multiplexed DNA sequence capture of mitochondrial genomes using PCR products. PLoS One 5(11):e14004
15. Noonan JP et al (2006) Sequencing and analysis of Neanderthal genomic DNA. Science 314:1113–1118
16. Sanger (2010) ftp://ftp.sanger.ac.uk/pub/pulldown/array20hyb20protocol.pdf
17. Sanger (2010) ftp://ftp.sanger.ac.uk/pub/pulldown/Solution20hyb20protocol.pdf
18. Meyer M, Stenzel U, Hofreiter M (2008) Parallel tagged sequencing on the 454 platform. Nat Protoc 3(2):267–278
19. Paul N, Yee J (2010) PCR incorporation of modified dNTPs: the substrate properties of biotinylated dNTPs. Biotechniques 48(4): 333–334

Chapter 22

Case Study: Enrichment of Ancient Mitochondrial DNA by Hybridization Capture*

Susanne Horn

Abstract

In ancient DNA studies focusing on estimating population histories, genetic markers are sequenced from a large number of samples belonging to the same species. Targeting loci of interest using traditional PCR can be time-consuming, in particular when samples are not well preserved and multiple overlapping fragments are required. Here, I describe the process of generating DNA libraries from ancient DNA (aDNA) extracts for high-throughput sequencing. I use a serial in-solution DNA hybridization approach with subsequent bead capture to enrich libraries for the target locus, in this case the mitochondrial control region of ancient beavers (*Castor fiber*). The resulting sequencing reads are run through quality control filters to obtain reliable consensus sequences. Using these sequences, I construct a phylogenetic tree, which agrees with previously published data regarding phylogeographic relationships among beavers.

Key words: Ancient DNA, Hybridization, Enrichment, High-throughput sequencing, Array capture, In-solution capture, *Castor fiber*

1. Introduction

Estimating the demographic history of ancient populations requires sequencing the same genetic locus from multiple ancient DNA extracts, which often vary considerably in the quality and quantity of preserved DNA. The workflow of this approach includes designing primers to amplify short, overlapping regions (around 100–200 base-pairs including priming sites), and replicating PCR amplifications

* *Note*: In the case study presented in this chapter, I describe the enrichment of target DNA from ancient DNA extracts using a hybridization-based method described in Chapter 21. I discuss specific challenges associated with using this method with ancient samples, including the generation of sufficient template DNA and the analysis of high-throughput sequencing data. For more information on the analysis of high-throughput sequencing data, see Chapter 23.

Beth Shapiro and Michael Hofreiter (eds.), *Ancient DNA: Methods and Protocols*, Methods in Molecular Biology, vol. 840, DOI 10.1007/978-1-61779-516-9_22, © Springer Science+Business Media, LLC 2012

to authenticate the resulting sequences and quantify DNA damage. This workflow is often hindered by the co-amplification of contaminating DNA molecules.

Within the last decade, considerable technical advances have made it possible to sequence entire ancient genomes (1–3). As many of these approaches require only a single DNA library to be prepared prior to sequencing, the amount of work required to generate large amounts of ancient DNA sequences is reduced considerably. However, whole genome sequencing remains prohibitively expensive for most labs, and provides genetic information for only a single individual. For population studies, a major advance has been to couple the targeted analysis of specific loci with new high-throughput sequencing technology. DNA hybridization capture makes it possible to target specific genomic regions from ancient DNA libraries. These regions can range in length from a few 100 base-pairs (bp) to megabases (Mb) in size.

Here, I describe the use of the hybridization capture protocol described in Chapter 21 (Horn, this volume) to isolate mitochondrial control region data from a sample of subfossil Eurasian beaver (*Castor fiber*) remains. Extant beavers are known to have pronounced phylogeographic structure (4). The goal of the project was to determine whether these phylogeographic patterns were already present in ancient beavers.

2. Methods

2.1. DNA Extraction and PCR Screening

I extracted DNA from 103 *C. fiber* bone and tooth samples ranging in age from 400 to approximately 45,000 years old, as described previously (5). I amplified two fragments of the mitochondrial control region, each around 90 bp in length including primers, to assess DNA preservation. Negative controls from the extractions did not yield PCR products of the expected size and therefore did not show signs of contaminating DNA. The 70 extracts that yielded at least one out of two PCR products were enriched for a 495-bp stretch of the mitochondrial control region using hybridization capture.

2.2. Library Preparation and Quantification

I prepared barcoded genomic libraries from 70 ancient DNA extracts and 8 negative controls for sequencing on the Illumina GAII (6). Four negative controls were included at the time of DNA extraction and another four at library preparation, both of which I performed in the clean room. I quantified the sequencing libraries using quantitative PCR (qPCR). Most samples yielded between 3×10^6 and 1×10^8 copies per microliter (cp/μL), whereas negative controls yielded around 1×10^5 cp/μL. The qPCR products were visualized on agarose gels to identify those libraries that contained inserts and those that contained only adapter dimers formed during library preparation, as may occur when the amount of extracted DNA is insufficient for library preparation (6, 7). The latter was the case for all negative controls. Libraries prepared from beaver samples that contained insufficient DNA were not processed further. Negative controls were carried through enrichment and sequencing in order to evaluate false assignment rates of barcodes and potential contamination at low levels (7).

2.3. Generation of Bait

I amplified a roughly 650 bp stretch of beaver mitochondrial control region, which was designed to overlap the target 495 bp fragment on both ends, using biotin-11-dUTP (5 μM final concentration) as described in Chapter 21. Including the flanking region in the bait molecules ensures that the entire 495 bp target region is captured efficiently. I used the purified amplification product as bait for hybridization capture, which I performed in 96-well plates.

2.4. Serial Hybridization Capture

I performed two serial hybridization captures of the genomic libraries using biotinylated bait as described in Chapter 21. The hybridization mixture contained about 17 ng of bait DNA and 170 ng of library in each well of a 96-well plate. For each hybridization capture, I placed the plate containing the hybridization mixture in a thermal cycler, heated it to 95°C for 5 min, and then cooled it to 65°C with 0.1°C/s, followed by incubation at 65°C for 24 h. I then immobilized the hybridized DNA using Dynabeads (Invitrogen) as described in Chapter 21.

After the first hybridization capture, I amplified the libraries using the Phusion® High Fidelity PCR master mix (Finnzymes) (6). I cleaned the reactions using the AMPure XP kit (Agencourt) and used them in a second round of hybridization capture. Performing capture twice increases the yield of mitochondrial control region molecules for sequencing. I then amplified and cleaned the resulting libraries using the Phusion® High Fidelity PCR master mix (Finnzymes) and the AMPure XP kit.

I then quantified the eluates containing the sequencing libraries enriched for mitochondrial control region DNA using a Nanodrop photospectrometer. This information was used to pool the libraries (both samples and negative controls) in equimolar amounts for sequencing.

2.5. Sequence Analyses

Illumina base calling was performed using the software IBIS (8) and sequencing reads were sorted according to their corresponding barcode sequence as described in Chapter 23 (6). I then mapped the sequencing reads to a control region sequence of *C. fiber* using the software bwa v0.5.5 (9). Reads were discarded unless they had a minimum mapping quality of 20 and a minimum length of 30 bp. Samples and negative controls for which fewer than 5% of reads mapped to the target region were discarded from further analysis.

Since I used PCR amplification of the library several times during the experiment, the same starting molecule may have been sequenced multiple times. For this data set, each read that mapped to the target locus was sequenced around 100 times on average. I therefore applied an additional filter in which these high frequency reads that start and end at the same position were only stored once (see Chapter 23). Low frequency reads were discarded because they often differed in their sequence from a high frequency read only by short indels. Thus, low frequency reads likely resulted from polymerase slippage, and were discarded prior to the generation of the consensus sequence. This was achieved by requiring that each high frequency read was observed 10 times at minimum. If more than 20 of the high frequency reads were present, they were used to create contigs of the target sequence. Finally, only contigs that covered more than 95% of the target were used for the generation of consensus sequences.

2.6. Phylogenetic Analyses

I aligned the consensus sequences to *Castor fiber* control region sequences from GenBank using ClustalW as implemented in BioEdit (10). I constructed a preliminary genealogy in Mega4 (11) using the neighbor-joining algorithm with the Kimura 2 parameter evolutionary model, the pairwise deletion criterion, and 1,000 bootstrap replicates.

3. Results

3.1. Sequence Analyses

The Illumina run yielded sequence data for all of the barcodes used, including those that were used for libraries created from negative controls. The raw number of sequencing reads per barcode reflects the relative pooling of all libraries, which may be influenced by pipetting and quantification errors. Therefore, the success of the enrichment and sequencing needs to be evaluated based on the percentage of reads that map to the target genomic region for each library. Enriched beaver sequencing libraries, on average, yielded around 24% (0.1–62.7%) of the reads mapping to the reference sequence. Six out of eight negative controls yielded fewer than 3% (0.7–2.8%) of reads mapping to target. Two negative controls yielded 12.2 and 16.8% of reads mapping to target, respectively; both of these had been included at the later step of library preparation.

Based on the counts for an unused index, the false assignment rate of indexes was estimated to be 1 in 6,400, similar to previously reported values ranging between 1 in 1,000 and 1 in 10,000 (7). After processing the sequencing reads through quality control filters described above, only sequencing libraries prepared from beaver samples remained for the construction of consensus sequences. Out of 70 ancient *Castor fiber* samples processed through enrichment and sequencing, 33 provided sufficient high-quality sequence data to reconstruct their mitochondrial control region sequences.

4. Discussion

DNA hybridization capture can be an efficient method for the targeted enrichment of many samples in parallel. To ensure that endogenous DNA survives in a sample prior to processing using this approach, it is recommended to screen the samples using PCR. While none of the negative controls met the applied quality control filter criteria, neither did about half of the beaver samples (33 of 70), suggesting that the pre-screening process employed in the first stages of this experiment was not sufficiently strict. In the cases of the less well-preserved specimens, conventional PCR may have succeeded in amplifying the target region despite the survival of only a few starting template molecules. Quantitative PCR may be used to improve the efficiency of the initial screening by discriminating well-preserved samples from poor samples. After sequencing, it is important to apply further quality filters to the data produced. The quality and length filters applied during mapping are useful to select sequences that originate from endogenous target-DNA. Filters to identify and account for high frequency reads are also useful to identify "real" sequences and generate the consensus sequence, in particular when the experimental setup includes amplification steps.

Even when applying these stringent filter criteria, considerable challenges remain in the analysis of high-throughput sequencing data. Very deep sequencing, such as the 100× coverage obtained on average here, may be more sensitive to recovering contaminating DNA molecules (12). This may explain why two out of eight negative control sequencing libraries initially (prior to applying quality control filters) showed more than 10% of sequencing reads mapping to the targeted genomic region.

In the classic approaches of targeted aDNA research, as soon as negative controls prove to be PCR negative, they are excluded from downstream analyses such as cloning and sequencing. Here, negative controls were carried throughout the entire experiment including sequencing. Quantitative PCR results suggest that the negative control sequencing libraries contained very low copy numbers of sequence, and agarose gel analyses suggested they were insert-free. However, the deep sequencing results showed low levels of sequence in two of the negative control libraries. As it is unclear when this contamination was introduced to the negative controls, this underscores the importance of using extreme care when handling all samples simultaneously in a 96-well plate for enrichment and amplification.

In addition to sequences in the negative controls, sequences were observed that mapped to the target region but also to an unused barcode. This most likely reflects sequencing error (13). The quality control filter that selects for multiply-amplified molecules may help to alleviate this problem.

4.1. Phylogeography

A preliminary phylogenetic tree comprising a subset of the *Castor fiber* control region sequences obtained in this experiment is shown in Fig. 1. The results support the previously recognized western and eastern groups of *Castor fiber* (4) providing further evidence for the authenticity of the sequences. DNA capture by hybridization

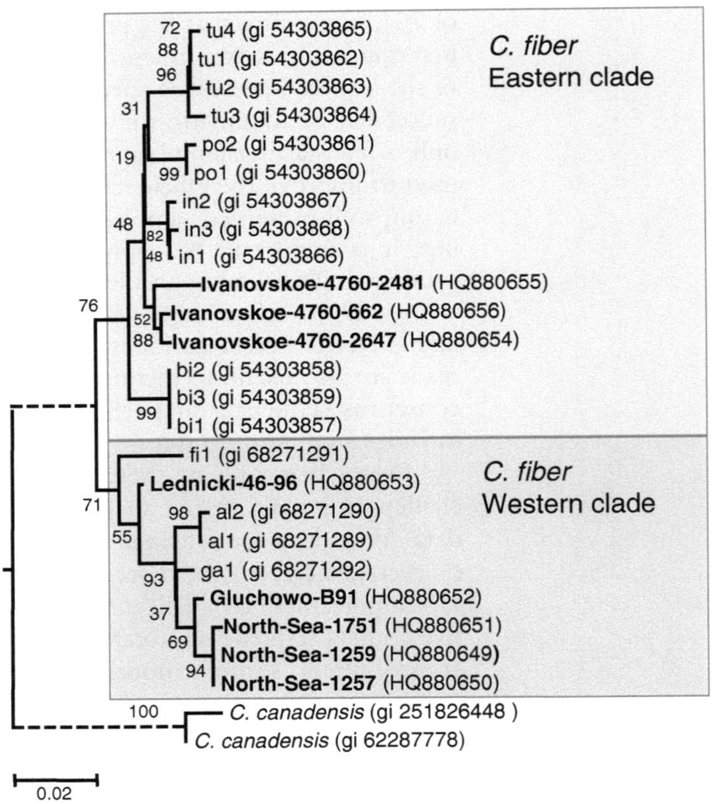

Fig. 1. Genealogy of mitochondrial control region sequences of ancient and extant Eurasian beaver, *Castor fiber*. Ancient beavers from Europe clustered into two groups: western beavers and eastern beavers. Accession numbers are given in brackets. The tree depicted is a neighbor-joining tree based on a 495-bp alignment (including gaps) rooted with sequences of the North American beaver *Castor canadensis*. Bootstrap support values are shown at the nodes.

is able to facilitate large-scale studies on phylogeography and genetics of ancient populations. The utilized protocol allowed the custom design of bait molecules in a standard molecular lab setting. This method is associated with lower costs than array capture and can be seen as an alternative when shorter loci, up to kilobases in length, are to be enriched from ancient DNA.

Acknowledgments

I would like to thank the Volkswagen foundation and Max Planck society for funding, M Meyer, the sequencing group and the bioinformatics group at the MPI EVA for their support in high-throughput sequencing, M Kircher for help in sequence analysis, and M Stiller for critical reading of this manuscript. C Schouwenburg, D Makowiecki, and T Kuznetsova provided the beaver samples shown in Fig. 1.

References

1. Rasmussen M et al (2010) Ancient human genome sequence of an extinct Palaeo-Eskimo. Nature 463(7282):757–762
2. Green RE et al (2006) Analysis of one million base pairs of Neanderthal DNA. Nature 444(7117):330–336
3. Reich D et al (2010) Genetic history of an archaic hominin group from Denisova Cave in Siberia. Nature 468(7327):1053–1060
4. Durka W et al (2005) Mitochondrial phylogeography of the Eurasian beaver *Castor fiber* L. Mol Ecol 14(12):3843–3856
5. Rohland N, Hofreiter M (2007) Ancient DNA extraction from bones and teeth. Nat Protoc 2(7):1756–1762
6. Meyer M, Kircher M (2010) Illumina sequencing library preparation for highly multiplexed target capture and sequencing. Cold Spring Harb Protoc 2010(6):pdb.prot5448. doi:10.1101/pdb.prot5448
7. Meyer M, Stenzel U, Hofreiter M (2008) Parallel tagged sequencing on the 454 platform. Nat Protoc 3(2):267–278
8. Kircher M, Stenzel U, Kelso J (2009) Improved base calling for the Illumina Genome Analyzer using machine learning strategies. Genome Biol 10(8):R83
9. Li H, Durbin R (2009) Fast and accurate short read alignment with Burrows-Wheeler transform. Bioinformatics 25(14):1754–1760
10. Hall TA (1999) BioEdit: a user-friendly biological sequence alignment editor and analysis program for Windows 95/98/NT. Nucleic Acids Symp Ser 41:95–98
11. Kumar S, Tamura K, Nei M (2004) MEGA3: integrated software for molecular evolutionary genetics analysis and sequence alignment. Brief Bioinform 5(2):150–163
12. Kircher M, Kelso J (2010) High-throughput DNA sequencing—concepts and limitations. Bioessays 32(6):524–536
13. Dohm JC et al (2008) Substantial biases in ultra-short read data sets from high-throughput DNA sequencing. Nucleic Acids Res 36(16):e105

Chapter 23

Analysis of High-Throughput Ancient DNA Sequencing Data

Martin Kircher

Abstract

Advances in sequencing technologies have dramatically changed the field of ancient DNA (aDNA). It is now possible to generate an enormous quantity of aDNA sequence data both rapidly and inexpensively. As aDNA sequences are generally short in length, damaged, and at low copy number relative to coextracted environmental DNA, high-throughput approaches offer a tremendous advantage over traditional sequencing approaches in that they enable a complete characterization of an aDNA extract. However, the particular qualities of aDNA also present specific limitations that require careful consideration in data analysis. For example, results of high-throughout analyses of aDNA libraries may include chimeric sequences, sequencing error and artifacts, damage, and alignment ambiguities due to the short read lengths. Here, I describe typical primary data analysis workflows for high-throughput aDNA sequencing experiments, including (1) separation of individual samples in multiplex experiments; (2) removal of protocol-specific library artifacts; (3) trimming adapter sequences and merging paired-end sequencing data; (4) base quality score filtering or quality score propagation during data analysis; (5) identification of endogenous molecules from an environmental background; (6) quantification of contamination from other DNA sources; and (7) removal of clonal amplification products or the compilation of a consensus from clonal amplification products, and their exploitation for estimation of library complexity.

Key words: High-throughput sequencing, Next-generation sequencing, Illumina/Solexa, 454, SOLiD-barcode, Sample index, Adapters, Chimeric sequences, Quality scores, Endogenous DNA, Contamination, Ancient DNA

1. Introduction

The advent of high-throughput sequencing (HTS) technologies has dramatically changed the scope of ancient DNA (aDNA) research. Beginning with Roche's 454 instrument in 2005 (1), and quickly followed by technologies from Illumina (2), Life technologies (3) and other companies (4–6), it is now possible to generate gigabases of sequence data within only hours or days. Shotgun sequencing of aDNA extracts (7–9) or aDNA libraries that have

Beth Shapiro and Michael Hofreiter (eds.), *Ancient DNA: Methods and Protocols*, Methods in Molecular Biology, vol. 840,
DOI 10.1007/978-1-61779-516-9_23,

been enriched for specific loci (10–13) provide a new window into preserved genetic material. For example, the first high coverage mitochondrial genomes (14–16) made it possible to characterize DNA preservation, contamination, and damage (17–19) to an extent that had not been achieved previously. As the cost of sequencing continues to decrease (20, 21), it has become feasible to analyze entire genomes of ancient samples (7–9, 22), including those for which the endogenous DNA makes up only a very small percentage of the total DNA extracted (8).

While the application of HTS to aDNA research is promising, the consequent increase in the amount of sequence data produced presents another challenge: efficient and reliable data postprocessing. Rather than aligning individual sequencing reads, millions of short reads, generally between ~36 and 300 nucleotides (nt) in length, must be analyzed. The highly fragmented nature of aDNA is ideal for such short-read technologies. However, other characteristics of aDNA extracts, including postmortem damage, the presence of coextracted DNA from environmental and other contaminants, and, often, a large evolutionary distance between the ancient sample and its closest modern reference sequence (23), can be problematic. These and various platform-specific problems can lead to substantial variation in run quality, high error rates, and adapter/chimera sequences, all of which confound assembly and analysis (20, 24, 25, 26).

Here, I outline typical primary data analysis workflows for aDNA experiments using HTS. I describe seven specific bioinformatics workflows: (1) separation of individual samples in multiplex experiments; (2) removal of protocol-specific library artifacts; (3) trimming adapter sequences and merging paired-end sequencing data; (4) base quality score filtering or quality score propagation during data analysis; (5) identification of endogenous molecules from an environmental background; (6) quantification of contamination from other DNA sources; and (7) removal of clonal amplification products and/or the compilation of a consensus from clonal amplification products, and their exploitation for estimating library complexity.

The workflows described below assume a paired-end Illumina Genome Analyzer data set, e.g., see refs. (8, 11, 27, 28). However, the instructions provided should apply to most types of HTS data sets generated from aDNA libraries.

2. Materials

2.1. Output Files from the Sequencing Platform

Each HTS platform, and, often, different versions of the same platform, produces slightly different output files. Non-vendor software packages are available that claim to improve the data quality of the original output file (29–33), and these, too, may produce a different type of output file. To simplify presentation of a bioinformatics workflow that is generalizable across platforms, I assume that the

```
...
@SOLEXA-GA03_0001_PEi_SG:5:1:1033:5267
AGACAGACACAGAGNAAGACCCAGTCCGCCACACAGGCAAACTCACGCAGTACGCGCCG
+SOLEXA-GA03_0001_PEi_SG:5:1:1033:5267
4--'-(/.23/044!51/+//.400/-/1-62/.6021834///62126313/2230+2
...
```

Fig. 1. A typical FastQ File. A FastQ file begins with "@" followed by a unique sequence identifier (these are platform specific; shown is an Illumina Genome Analyzer read ID providing run ID, lane, tile, and *X*–*Y*-coordinates). The next line contains the sequence. The third line begins with "+" and can be followed again by the complete read identifier. The fourth line contains the quality scores. This example encodes quality scores using the Sanger standard (see Note 3) with ASCII characters from 33 to 127 encoding base qualities in PHRED-scale between 0 and 94 (e.g., "4" (ASCII 52) corresponds to PHRED quality score 19 and thus an error likelihood of 1.26%, while "-" (ASCII 45) corresponds to PHRED 12 and an error probability of 6.31%).

user will begin with a file of nucleotide sequence data with a quality score associated with every base.

The most common file format used both in data exchange and as input for post-sequencing software is currently the FastQ format. FastQ is an extension of the FASTA sequence format, where each sequence in the file is associated with an identifying tag and with an additional line for quality scores (see Fig. 1). Depending on which HTS platform is used, it may be necessary to convert output files to FastQ format for further processing (see Notes 1 and 2). Unfortunately, there is no universally accepted rule regarding how quality scores are encoded. It is generally recommended to follow the Sanger standard (one character per quality score with an offset of 33, see Note 3), as this is currently the most widely used format. Always consult the documentation and default parameters of the software you use for specific requirements of the FastQ input files.

2.2. Hardware

Given the large amount of data generated by HTS, the FastQ sequence files can be very large: each may contain several million sequence reads, and 4 times as many lines. The efficient processing of these gigabyte-sized text files requires access to computational resources that typically exceed normal desktop computers (minimum requirements: 4–8 cores, 16–32 GB of memory, ~500 GB disk space for intermediate and output files). Due to the large file sizes, it is advisable to store compressed versions of these files in order to reduce input/output operation bottlenecks on network and local file systems.

2.3. Software

Most software currently available for data processing runs on UNIX-based systems such as Linux and Mac OS, but may also work in a Windows cygwin environment (34). Python, bash, or Perl scripts can be efficient for linear extraction of information from text files and for writing to intermediate and output text files. Available bioinformatics packages (e.g., bioperl (35) and biopython (36)) provide useful functions and data structures for working with FastQ and sequence data in general. Large amounts of sequencing data may, in some applications, create the need for more efficient and indexed data storage (e.g., bioHDF (37), Google's BigTable (38), Apache Hadoop (39)).

The example workflows provided below assume a Linux operating system, in which the necessary programs/scripts can be called from a central installation.

2.4. Programs and Scripts

List of programs and scripts that will be required in the protocols described below.

BWA (v0.5.8a)	http://bio-bwa.sourceforge.net
cdhit	http://weizhong-lab.ucsd.edu/softwares/cd-hit-454/cd-hit-454.tar.gz
FastQC	http://www.bioinformatics.bbsrc.ac.uk/projects/fastqc/
KeyAdapterTrimFastQ_cc, QualityFilterFastQ.py, SplitMerged2Bwa.py, ContTestBWA.py, FilterUniqueSAMCons.py, SplitMerged2CDhit.py, SplitFastQdoubleIndex.py, TrimFastQ.py, MergeReadsFastQ_cc	http://bioinf.eva.mpg.de/fastqProcessing/
Python	http://www.python.org/download/
R	http://cran.r-project.org/
samtools (v0.1.7a)	http://samtools.sourceforge.net
TagDust	http://genome.gsc.riken.jp/osc/english/dataresource/

3. Methods

In many of the examples below, simple commands need to be typed into the command-line interface of your computer. In most cases, I first provide descriptions, in words, of what the command is expected to achieve. I then provide (in bold face font) the actual commands that you should type in order to achieve the expected results.

3.1. Separation of Individual Probes in Multiplex Experiments

Multiplexing and sample pooling are becoming more common in HTS experiments. Barcoding (also called indexing or tagging) is often used when the target sequences comprise only a few loci or small genomes, and therefore sequencing only a single individual per lane or region would yield an excessive coverage. While sequencing platforms differ in how barcoded libraries are constructed, sequences from different libraries will be computationally sorted post-sequencing via the index that is either part of the actual sequence read (index adjacent to insert) or sequenced as a separate technical read (index embedded in the adapter sequence). Typically, authors of the different barcoding protocols also provide software for separation by index (demultiplexing, e.g., see refs. (40–43)) with the common result that the pooled sequences from a single run are written to separate files based on the identified index sequence.

Demultiplex approaches differ in (1) whether only exact matches from a list of used/available index sequences are identified or whether a limited number of errors is allowed, and (2) whether quality scores in the barcode read are used in this process. While using exact matches and requiring high quality scores provide the most conservative approach, this may not always be feasible. With long barcodes, for example, sequencing error may cause too many sequences to be excluded. Assuming a uniform error rate of 0.5% and a 6 nt index, around 3% of reads are predicted to contain errors. When barcode length is increased to 10 nt, around 5% of reads are expected to contain errors. Imbalanced use of barcodes (40) and the nonuniform distribution of errors across sequences (21, 25) will increase the proportion of erroneous reads (~5–15%) with some sequence readouts showing close to random sequence. An intermediate solution is to apply a quality filter (see Subheading 3.4) and to match a set of precompiled index variants containing very few substitutions.

Below, I provide an example workflow to separate sequences based on their index nucleotides. I use raw FastQ data from a 2×101-cycle Illumina paired-end sequencing lane ("s_8_sequence.txt"). This data set contains a pool of 96 samples and an indexed ϕX174 sequencing control library. All sample libraries have an average insert size of less than 200 nt. Half of the samples originate from ancient specimens while the other half are from modern specimens. In this experiment, indexed ancient and modern DNA samples have been pooled and used in an enrichment procedure (for mitochondrial sequences); therefore, the number of reads associated with each sample is expected to vary considerably. Samples are identified by two, 7 nt indexes that are read in two technical reads: one after the forward read and the other after the reverse read. The FastQ file therefore contains sequences with a length of 216 nt: 101 nt forward read, 7 nt first index, 101 nt reverse read, and 7 nt second index. The multiplex approach is an extension of the protocol described by Meyer and Kircher (40), in which the IS4 primer is replaced by a set of different index primers that introduce the second index read (44). This setup is useful for identifying and excluding experimental artifacts, such as index contamination or jumping PCR (45–48) that may occur during pooled library amplification.

To create separate FastQ files for each index, use the SplitFastQdoubleIndex Python script. In addition to the original FastQ file, provide the program with a three column, tab-separated text file with first and second index sequence followed by the name of the sample (see Note 4). The file may contain one header line introduced by the hash character ("#"). For example:

```
==== samples.tsv ====
#SEQ SEQ2 NAME
ACAGTGA   CTCGATG   che:arc
AGGCCGA   ACCAACT   che:pha
TGCTGGA   TTGAAGT   che:mrz
[...]
AGATTCT   GCAAGAT   tar:mes
CAACTAA   AACTCCG   tar:var
TTGCCGC   AGATCTC   control
```

When analyzing libraries created with either the original Meyer and Kircher (40) protocol or similar protocols (42, 49), the same script can be used but with a two column file, leaving out the second index sequence. The script allows the user to choose whether to require perfect matches to each index or to allow mismatches between the sequence reads and the index used (see ref. (40) for details).

When using error correction, there can be either up to one mismatch or the loss of the first base between the index that is read from the sequences and the "true" index sequence provided in the sample file. The script will create a separate output file for each of the defined index pairs and for index readouts that are found in the output FastQ files but that have not been identified by the user's sample file ("unknown" indexes). In addition, a file will be created for index variants that cause two defined indexes to be less than two mismatches apart from each other ("conflict" indexes), so that it is not possible to determine with certainty to which sample the index belongs. If index pairs are used for sample identification and combinations of indexes encountered that are not defined by the user (incorrectly paired indexes), these reads are written into a "wrong" index file. Finally, valid index reads that have read quality scores below a provided threshold value can be automatically excluded. These quality-excluded sequences are saved in the "unknown" file with an asterisk ("*") added to the FastQ read identifier.

In the following example, run the demultiplex script on the example data set. First, using the command-line interface of your computer, create a new folder for the output files by typing the following command at the prompt:

```
$ mkdir split
```

Next, call the SplitFastQdoubleIndex script with the file containing the expected index pairs (*samples.tsv*). Define the output folder and the start of the paired-end read in the FastQ input file (*s_8_sequence.txt*). Require a minimum quality score of 10 for the index reads and ask for a summary when the script finishes:

```
$ SplitFastQdoubleIndex.py -i samples.tsv -o split/ --
start=109 \
  --quality=10 --summary s_8_sequence.txt
Error: Skipping first base in index causes a conflict
GCTCGAA->CTCGAAG with variant of CTCGATG
Summary:
  Observed    4208449 tags for unknown (unknown)
  Observed    2564524 tags for lem:pha (('ATCTATA',
'ACCAACT'))
  Observed    1734817 tags for tar:mic (('TGCCATA',
'CCGGTAC'))
  Observed    1550220 tags for wrong (wrong)
  Observed    1166226 tags for lor:pha (('CATTTTA',
'ACCAACT'))
  Observed    1036767 tags for tar:var (('CAACTAA',
'AACTCCG'))
  Observed    1019546 tags for pro:pha (('ATGAGCA',
'ACCAACT'))
[...]
  Observed      49979 tags for lem:meg (('GAGTGGA',
'TCGCAGG'))
  Observed      43080 tags for lem:pac (('AACGAAC',
'CTCTGCA'))
  Observed      19677 tags for pro:ar2 (('ACGCGGC',
'GGATCAA'))
  Observed       7437 tags for conflict (None)
```

When changing to the output folder (split), you should now see 99 output files: one for each sample (96) plus the conflict, unknown, and wrong files as described above. The new FastQ sequence files no longer contain the sequence and quality scores of the index. Thus, sequences in the output files have a length 202 nt: 101 nt forward read and 101 nt reverse read.

```
$ cd split
split$ ls s_8_*
s_8_che:ar1_sequence.txt    s_8_che:ar2_sequence.txt
s_8_che:arc_sequence.txt    s_8_che:euo_sequence.txt
[...]
s_8_unknown_sequence.txt    s_8_wrong_sequence.txt
```

Of the 47,584,117 sequences in *s_8_sequence.txt*, 4,208,449 (8.84%) are assigned to "unknown" indexes (931,550 or 1.96% of these were excluded by using a minimum quality score cutoff of 10). Further, 1,550,220 (3.26%) sequences are "wrong" pairs and 7,437 are "conflict" (0.02%; this is because two of the second read indexes are not at least two bases distant from each other if the loss of the first base of the index is allowed). For the 96 samples used in this example, we obtain an average of 432,264 sequences per sample. As expected, the variation between samples is large: the best represented sample has 130 times more sequences than the least represented sample (minimum 19,677; maximum 2,564,524). In an ideal multiplex experiment, this factor should not exceed 10.

3.2. Removal of Protocol-Specific Library Artifacts

Most HTS platforms require platform-specific adaptors to be ligated to the molecules in the DNA libraries prior to sequencing. Library preparation protocols vary in their propensity to create library adapter dimers, chimeric sequences, and other artifacts that will need to be identified and removed. In a typical HTS experiment using modern DNA, protocols are followed to enrich for molecules with correctly added adapters and to remove molecules with short or no inserts. When this is not possible, as is often the case with aDNA, library artifacts may dominate the resulting sequencing reads.

Programs such as TagDust (26) compare the original adapter and primer oligonucleotide sequences with the output files to identify artifacts. The program can be used either to remove all sequence for which the library preparation oligonucleotide *k*-mers (see Note 5) make up the majority of the sequencing read (direct filtration) or to cluster the results from a single lane and identify the most frequently observed sequences, which can then be used for trimming and filtering with other software (see Subheading 3.3). If TagDust is used to filter aDNA sequencing data, reads of short insert size (as may be common in degraded samples) may be excluded, simply because a large part of the sequence comprises library preparation oligonucleotides rather than the insert sequence. This could remove potentially informative sequences from the analysis.

TagDust requires a FastQ file with single read data. Below, use only the first 50 nt of the forward read to identify artifacts. Generate the files using the TrimFastQ.py script. In addition, provide a list of the adapter sequences in a single FASTA file formatted as in the example file *double_multiplex_adapter.txt* below:

```
==== double_multiplex_adapter.txt ====
>P5graft
AATGATACGGCGACCACCGAGATCTACAC
>GenomicR1
ACACTCTTTCCCTACACGACGCTCTTCCGATCT
>GenomicR2
AGATCGGAAGAGCACACGTCTGAACTCCAGTCAC
>P7graft
ATCTCGTATGCCGTCTTCTGCTTG
```

Run TrimFastQ.py, to create the input file with all sequences trimmed back to the first 50 nt of the forward read:

```
$ TrimFastQ.py -e 50 s_8_sequence.txt
```

Call TagDust, and request to write the identified artifacts in FASTA format to a file called *artifacts.fa*. Provide the FASTA file with the adapter sequences (*double_multiplex_adapter.txt*) as well as the trimmed reads in FastQ:

```
$ tagdust -fasta -a artifacts.fa
double_multiplex_adapter.txt \
 s_8_sequence_trim.txt
```

Obtain the most frequent sequences in *artifacts.fa* using GNU command line tools. The following command line ignores all lines in *artifacts.fa* that look like a FASTA header (i.e., start with ">"), sorts the remaining sequence lines, counts how many different sequences are obtained, then sorts them reverse numerical order and prints out the 30 most frequent sequences:

```
$ grep -v "^>" artifacts.fa | sort | uniq -c | sort -nr |
head -n 30
   2443 ACACTCTTTCCCTACACGTCTGAACTCCAGTCACACTCGTTATCTCGTAT
   1974 GGATTTCTCCGGTCTGAACTCAGATCACGTAGGACTTTAATCGTTGAACA
   1671 ACACTCTTTCCCTACACGTCTGAACTCCAGTCACAGAGACCATCTCGTAT
   1517 GAGCACACGTCTGAACTCCAGTCACACTCGTTATCTCGTATGCCGTCTTC
[...]
    441 ATTTCTCCGGTCTGAACTCAGATCACGTAGGACTTTAATCGTTGAACAAA
```

Now provide the same command as above, this time trimming the sequences so that only the first 30 nt of the identified artifact sequences are considered:

```
$ grep -v "^>" artifacts.fa | cut -c 1-30 | sort | uniq -c
| \
 sort -nr | head -n 30
  13646 ACACTCTTTCCCTACACGTCTGAACTCCAG
   5493 CGTATGCCGTCTTCTGCTTGAAAAAAAAAA
   2859 GAAGAGCACACGTCTGAACTCCAGTCACAG
   2332 GAAGAGCACACGTCTGAACTCCAGTCACAC
[...]
    671 AGAAAGAAAGAAAGAAAGAAAGAAAGAAAG
    659 GAGCACACGTCTGAACTCCAGTCACACTGC
```

Two distinct populations of sequences can be seen within the most frequent 30 sequences identified by TagDust: one matching adapter dimer variants and the other matching mitochondrial sequences (see Fig. 2). The second population likely results from the enrichment process, which may alter the *k*-mer representation in the

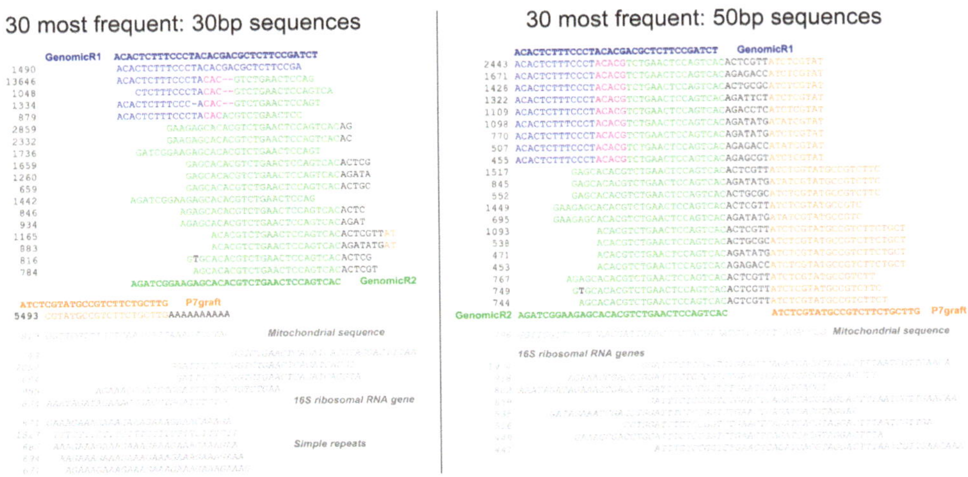

Fig. 2. Adapter dimer variants and a population of false positive sequences (*italic*) that match the 30 most frequent artifact sequences reported by TagDust (26). Here, I consider only the first 30 nt (*left*) or 50 nt (*right*) of the forward read for analysis.

samples and cause TagDust to identify as artifacts real sequences that happen to be present in high numbers. Thus, this direct application of computational filtering has its limits. In the next section, I will provide a protocol for filtering the identified adapter dimer variants. Other artifacts arising from the sequencing process are platform-specific and remain mostly unidentified using the approach outlined here (see ref. (20) and see Note 6 for further details).

3.3. Adapter Trimming and Merging of Paired-End Sequencing Data

Reads of short insert-size molecules often contain parts of the adapter sequence at the read end, which need to be identified and removed so as not to interfere with downstream mapping or alignment. Unfortunately, this is not part of most data processing pipelines (with the exception of 454/Roche, see Note 7). This step is also nontrivial when sequencing error probability increases toward the end of reads.

Adapter identification is simpler with paired reads than with single reads (8, 27, 28). For sequencing technologies without insertion/deletion errors, aligning the paired reads and identifying the overlapping region (autocorrelation) can reveal where the insert ends and adapter sequence begins (see Fig. 3). This approach is more powerful than any alignment (-like) process for identifying adapters in single reads, as these will frequently remove short pieces of non-adapter sequences or miss adapter sequences due to higher sequencing error rates at the ends of reads.

Merging paired-end reads in short-insert libraries also decreases the number of sequencing errors (28). On a simulated Illumina data set with uniform read length distribution, the merging approach applied in refs. (8, 11, 13, 27, 28) reduced the error rate of all merged sequences by a factor of about 5, and, for sequences shorter than or equal to the read length, by a factor of about 21

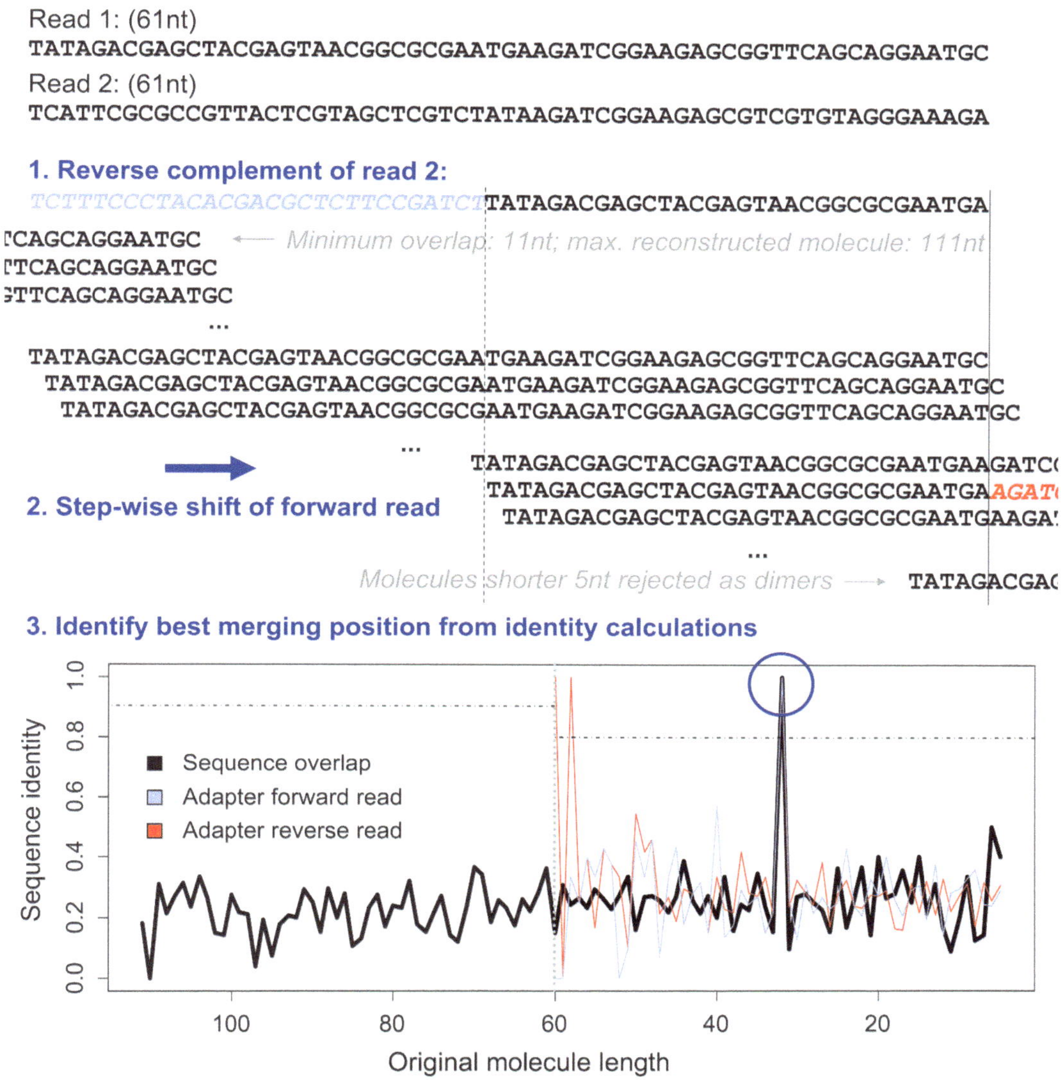

Fig. 3. Manipulating paired-end data. For paired-end data, identifying the adapter set-in point is simplified by searching for overlapping sequence shared by the forward and reverse read. The figure illustrates how the forward read is shifted along the reverse complement of the second read to identify the original molecule length and find the adapters (steps 1 and 2). This is similar to the approach applied in refs. (8, 11, 13, 27, 28), except: (1) the calculated sequence identity is corrected for the observed quality scores; and (2) a heuristic is implemented by first searching the variants of decreasing length with adapter sequence present, and then checking the longer variants with no adapter sequence by increasing length. The search is aborted when a sequence identity of 95% is observed (step 3). If 95% identity is not observed, the maximum sequence identity is considered for read merging (reads are merged if at least 90% identity is observed when no adapter is present, or 80% identity is observed when at least one of the adapters is present). The implementation requires a minimum length of 11 nt for the overlap and rejects inserts shorter than 5 nt as adapter dimers.

(see Fig. 4). Further, for sequences shorter than or equal to the read length, 99.997% of the simulated sequences were correctly merged. For the complete merging length range (5–191 nt) the original molecule length was recovered from 99.664% of reads. The most frequent reason for the merge failure was sequencing

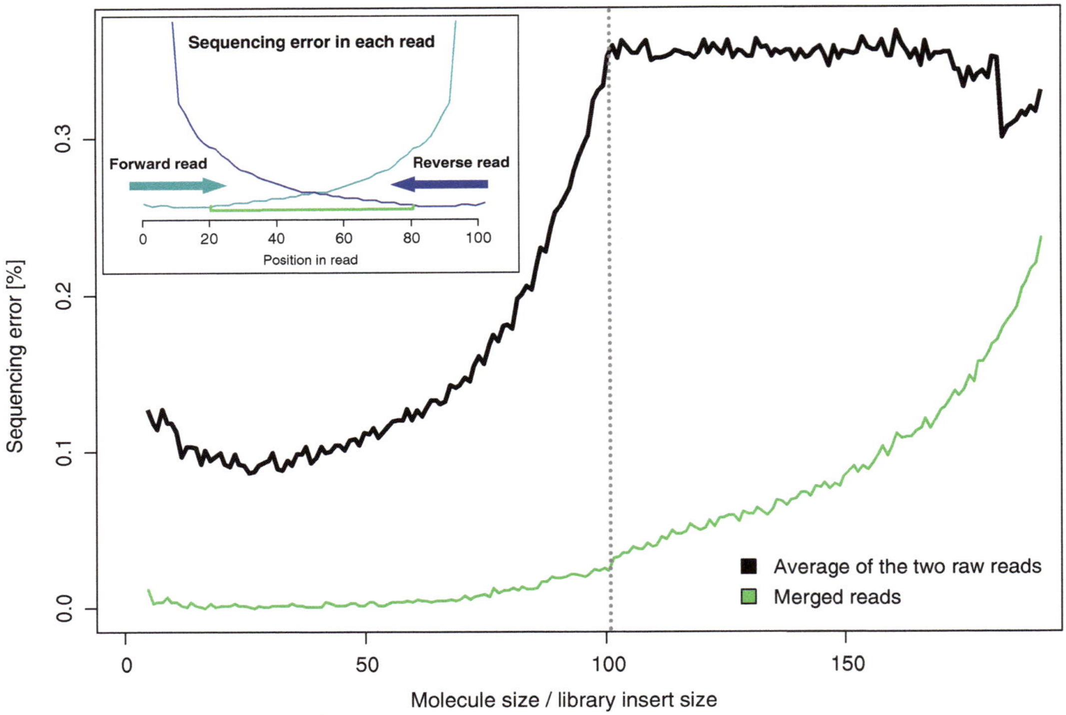

Fig. 4. Reduction in sequencing error rate caused by merging paired-end sequence data. Merging paired sequencing reads (*inset box* illustrates sequencing error of paired end reads) allows library adapters to be identified and removed efficiently and increases read accuracy. The *thick black line* shows the amount of sequencing error for different molecule lengths. Note that as the length of the read increases, so does the sequencing error. Alternatively, the *thin line* shows the sequencing error that remains after application of the read-merging algorithm described in the main text. The data depicted are from a simulated data set of sequences ranging in length from 5 to 191 nt, generated with 2 × 101-cycles on an Illumina instrument with v4 chemistry. The data set was simulated with an error-informative quality score for which a random number (between 0 and 10, uniform sampling) was added to the average quality score for this sequence position when the correct base was simulated, and a random number subtracted if a wrong base was simulated.

errors in the overlapping regions (0.259%); false merging results are reported in only 0.077% of cases and mostly trace back to simple repeat sequences. Merging long insert libraries, where the overlapping region may be very small or not present, may cause incorrect reconstruction in particular of repeat regions. In this simulated data set, on average 0.29% of longer sequences (192–350 nt) were incorrectly reported as merged reads.

If paired-end data is not available, a requirement can be imposed so that at least 5 nt of adapter sequence must be identified if a sequence is to be included in downstream applications (false adapter identification from 5 nt of random sequence is <0.1%). In addition, longer sequences (which may contain the adapter) can be excluded from all downstream analyses, so that only a minimal fraction of erroneous adapter sequences will be propagated into alignment and data analysis. Alternatively, read trimming can be combined with alignment, for example by allowing the alignment of the 3′ end to be extended either into the genome or the adapter. For short adapter pieces, the adapter and the reference may provide equally good alignment scores. In these cases, the alignment should be terminated and a trimmed alignment reported.

Since the example data set is paired-end, we will use the MergeReadsFastQ_cc program; for single read data, KeyAdapter TrimFastQ_cc is available. Any adapter chimeras that have been identified (see Subheading 3.2) can be provided to either program to be filtered from the output files.

Applying this to the example data set, first create a new subfolder *merge*:

```
split$ mkdir merge
```

Then call MergeReadsFastQ_cc in a for-loop for each of the files obtained from the demultiplex step (see Subheading 3.1). The *c* parameter defines a comma separated list of artifact sequences (obtained in Subheading 3.2), *f* defines the adapter starting at the end of the forward read and *s* the adapter starting at the end of the second read. Parameter *r* defines the start position of the second read in the provided sequence file, *o* the output folder, and *$i* is the iterator variable of the for-loop. Redirect the command line output to a log file called *merge.log*:

```
split$ (for i in s_8_*_sequence.txt; do MergeReadsFastQ_cc
\
 -c 'ACACTCTTTCCCTACACGTCTGAACTCCAG,\
ACACTCTTTCCCACACGTCTGAACTCCAGT,ACACTCTTTCCCTACACACGTCTGAACT
CC,\
CTCTTTCCCTACACGTCTGAACTCCAGTCA,GAAGAGCACACGTCTGAACTCCAGTCAC
II,\
GAGCACACGTCTGAACTCCAGTCACIIIII,GATCGGAAGAGCACACGTCTGAACTCCA
GT,\
AGATCGGAAGAGCACACGTCTGAACTCCAG,AGAGCACACGTCTGAACTCCAGTCACII
II,\
ACACGTCTGAACTCCAGTCACIIIIIIIAT,GTGCACACGTCTGAACTCCAGTCACIII
II,\
AGCACACGTCTGAACTCCAGTCACIIIIII,CGTATGCCGTCTTCTGCTTGAAAAAAAA
AA' \
 -f 'AGATCGGAAGAGCACACGTCTGAACTCCAGTCAC'\
 -s  'AGATCGGAAGAGCGTCGTGTAGGGAAAGAGTGTA'  -r  102  -o  merge/
$i; done) \
 >& merge.log
```

The merging script above will separate the two paired-end reads in the output file if merging is not successful. Merged reads can be identified in the output file by an "M_" at the beginning of the read identifier, while "F_"/"R_" indicate the forward and reverse read, respectively. The number of adapter dimers removed and the number of merged/non-merged reads are printed to command line when the script finishes. For the example data set, the fraction of removed adapter dimers/chimeras is on average 0.22% (minimum 0.00%; maximum 2.24%). If average numbers above 3–5% are observed, optimization of the library preparation protocol should be considered to improve future sequencing experiments. In the resulting files, 85.0% of reads are merged (minimum 61.8%; maximum 95.4%), which is typical for short insert sizes. By contrast, the ϕX174 control library, which has a considerably higher average insert size of ~350 nt, shows only 1.7% merged reads.

The length distribution of the merged reads provides an approximation of the length distribution of molecules in the aDNA library. We can obtain this distribution by extracting from the FastQ files all reads where the identifier starts with an "M_" and determining their length. The length counts can be plotted using the statistical programming language R (50). Figure 5 shows the first nine plots. The following command lines were used to generate them.

In a for-loop, iterate over all merged files. Use the command line tool awk to count the different sequence lengths observed for reads with read identifiers starting with "M_". Sort the output by number and write it to one file per sample. The output files have the extension ".length" instead of the extension ".txt" used for the FastQ files:

```
merge$ for i in *_merged.txt; do \
 cat $i | awk '{if (state) lenarray[length($1)]++;\
   if (/^@M_/) {state = 1} else {state = 0} }\
   END{for (len in lenarray)
printf("%5d\t%9d\n",len,lenarray[len])}'\
 | sort -n > ${i/.txt/.length}; done
```

Open the statistical programming language R (50), generate an output PDF *length_dist.pdf*, obtain a list of all *.length* in the current folder, iterate over these files, and plot them into the PDF:

```
merge$ R --vanilla --quiet
> pdf('length_dist.pdf',width=8,height=8)
> files <- list.files(pattern=".*.length$")
> for (filename in files) {
    data <- read.table(filename,header=F,sep="\t",as.is=T);

plot(data$V2~data$V1,type="l",xlab="Length",ylab="Count",
    main=filename,xlim=c(0,200)); }
> dev.off()
```

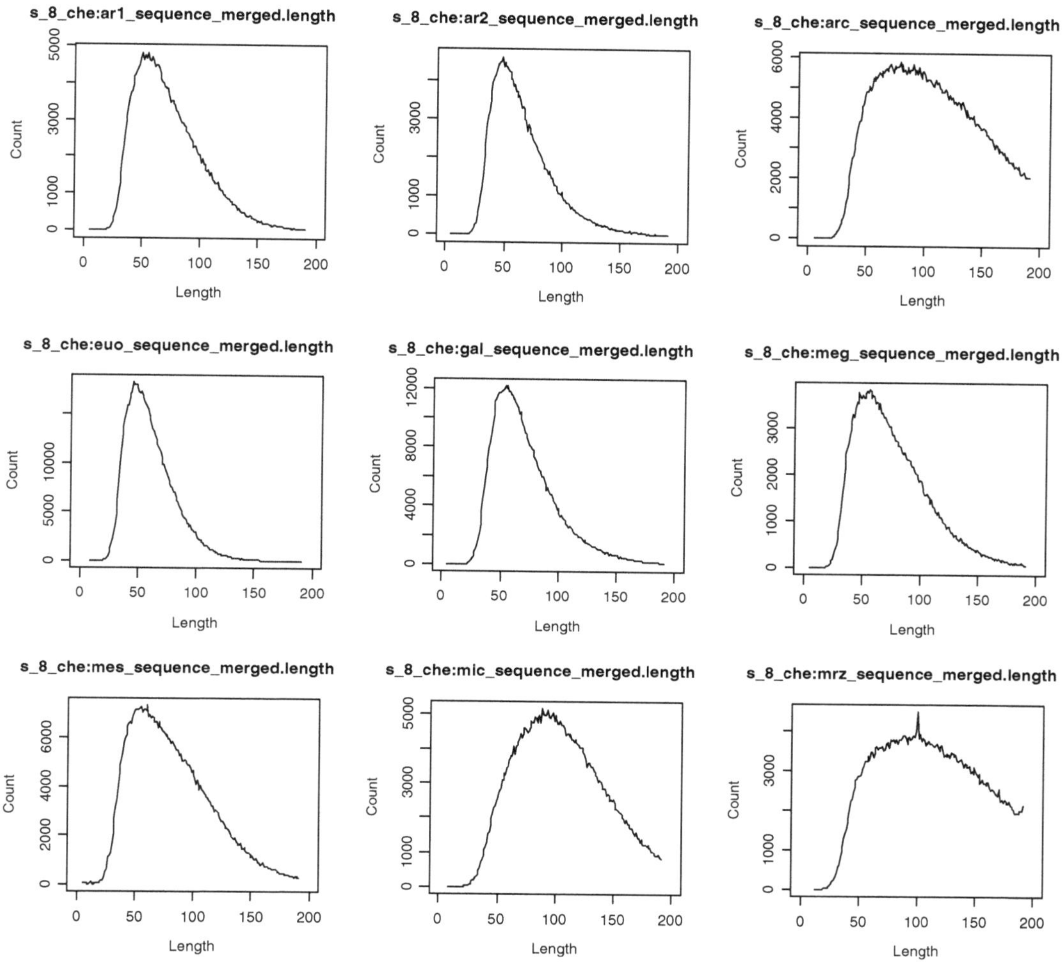

Fig. 5. Length distribution extracted from merged reads from 9 of the 96 samples. While some samples show very narrow length distributions (e.g., che:ar2, che:euo), others have wide distributions (e.g., che:arc, che:mrz, che:mic). The last plot (che:mrz) shows a higher number of sequences exactly at read length. This typically results from simple repeat sequences being merged incorrectly.

3.4. Quality Control and Filtering

Most HTS technologies immobilize the sequencing template using a random dispersal process. This results in different inter-sequence distances (either on a flat solid phase flow cell or in the form of beads carrying multiple templates) and will cause different susceptibility to mixed-signal readouts. In the most extreme cases, these mixed signals may result in nearly random sequences with overall low base quality scores, or a sequence that is very similar to one of the original molecules but that contains low base quality scores or high error rates for some positions along the read. This distance effect on signal purity causes sequencing errors to be nonrandomly distributed among reads. Typically, instrument-specific preprocessing of the raw data removes these sequences (see Note 8). However, a simple quality-score-based filter, which is by design highly correlated with signal purity, can be applied to all bases of the read to assure data quality independently of a specific preprocessing pipeline.

Many analysis programs that are able to read FastQ as input do not use the quality score information in the actual algorithm. Quality scores on the PHRED scale range mostly from 10 to 40 (10–0.01% error probability, see ref. (51)) and error probabilities above 1% are typically too high to be ignored for statistical inferences from the data. While several alignment programs and most genotype callers use these quality scores in their score calculation, many sequence annotation, motif finding, RNA folding, and phylogenetic analysis programs do not. For these types of programs, reads may be filtered for appropriate quality, or bases of low quality may be masked, for example by replacing the low-quality base with an ambiguous (N) basecall.

Various tools are available to produce run quality statistics (e.g., TileQC (52) or FastQC (53)). In addition to these, quality score recalibration based on alignment can be used to compare and adjust quality scores from different sequencing runs and different technologies (but see Note 9). At present, the most widely used algorithm is part of the Genome Analysis Toolkit (GATK) (54), and is used, for example, in the 1,000 Genomes project. Sequencing control libraries provides a means to calibrate quality scores between reactions and sequencing platforms. For our example data set, the Ibis base caller (32) used the ϕX174 control library reads to train its statistical models and to adjust the quality score estimates.

As a next step, we will assess the quality score and base distributions using FastQC. FastQC generates an HTML report for a FastQ file. Prior to running FastQC for each sample separately, we will run the complete raw sequence data file (execute the command line `fastqc s_8_sequence.txt`). Individual samples may be assessed if something unusual is observed in the full data file.

Figure 6 shows parts of the HTML report output by FastQC. There are five critical results as indicated by cross-symbols in the report: (1) In the per base quality score plot, a few cycles show reduced average quality scores, which is a reasonable result of quality score calibration. (2 + 3) Per base sequence and GC content plots show spikes around the index reads, indicating an unbalanced base composition of the indexes selected. (4) Per read GC content differs from the theoretical distribution, which can be caused when larger pieces of the adapters are included in the sequence reads or due to GC bias from the enrichment process. (5) *K*-mer analysis identifies patterns of 5-mer overrepresentation. These patterns match simple repeat sequences (these are also observed in the TagDust output and likely an artifact of the enrichment process) and motifs from the adapter sequences, the latter indicating that some adapter sequence is included in the read. Hence, the FastQC report for the whole run did not show any surprising results.

In downstream analyses, we will use programs that make limited use of the quality scores. The mapping program BWA (55), for example, will only look at quality scores for prioritizing the search for non-perfect sequence matches. We therefore apply a simple, low stringency quality filter to our data that removes all reads that have more than five bases with quality scores less than 15:

Create a new subfolder *quality* in *merge*:

```
merge$ mkdir quality
```

Iterate over the output files from the merging script and call the QualityFilterFastQ script with the quality threshold (*-c 15*) and the number of bases allowed below this threshold (*-n 5*), define the output folder using the *o*-option, and redirect the command line output into a text file (*quality.log*):

```
merge$ (for i in s_8_*_sequence_merged.txt; do
QualityFilterFastQ.py \
-c 15 -n 5 -o quality/ $i; done) >& quality.log
```

This filter removed on average 4.8% of the reads for each of the 96 samples (minimum 1.4%, maximum 10.3%), and discarded 24.7% of the nonidentified and low index quality sequences in "unknown."

3.5. Identification of Endogenous Molecules

One of the biggest challenges for aDNA is to distinguish endogenous sequences from those of environmental contaminants. In this situation, it is useful to know whether the potentially contaminating sequences have features that distinguish them easily from the endogenous DNA. GC content and length-based filters are typically insufficient to distinguish endogenous from exogenous DNA. *K*-mer frequencies, in particular the presence of longer *k*-mers, may be more useful (see Note 5).

Alignment approaches in which the sequence reads are aligned to the closest available reference sequence are often used. These may fail when the reference genome and the sample are evolutionarily distant, resulting in multiple substitutions or large indels. In this case, an overlap extension step (e.g., see ref. (56)) can be implemented, in which the unaligned reads (those that have not matched the reference genome) are used to attempt to extend the aligned reads (those that match the reference genome) into incompletely covered regions. Ideally, alignment and extension steps should be performed iteratively; however, there is no guarantee that such a process will converge. An iterative approach is also useful with *k*-mer filters, where *k*-mers identified within all reads that pass a first *k*-mer filter are used for an iterative identification step. Alternatively, de Bruijn graphs (e.g., *see* refs. (57–60)) can be used to first build contigs from overlapping *k*-mers. These contigs can then be filtered. Alignments may outperform *k*-mer/de Bruijn graph approaches when samples are experimentally enriched for similarity to specific sequences, as the enrichment approaches will enrich for these same *k*-mers in the environmental DNA molecules.

Filters are generally applied in two ways: (1) negative/subtractive filtering, where reads identified by some criterion are removed from the read set; and (2) positive filtering, where only reads identified by some criterion are kept in the read set. Negative filtering may leave false sequences in the read set and remove highly conserved sequences, while positive filtering may miss divergent regions and include highly similar (e.g., conserved) false sequences.

The sequences in our example data set were enriched for their similarity to strepsirrhine primate mitochondrial genomes. We will therefore define endogenous sequences by aligning the reads to any of the 14 complete mitochondrial genomes currently available from NCBI:

Create a FASTA file called *strepsirrhini.fasta* that contains these 14 sequences. Use the program BWA (55) to perform a semi-global alignment (see Note 10) with gaps and substitutions, and store its output in BAM, the binary version of the Sequence Alignment/Map format (61). BWA expects either single read or paired-end data. Due to read merging and quality filtering, we have a mixture of both data types, and need to separate them prior to alignment. Use BWA with the parameters " -l 65536 -n 0.01 -o 2" which will turn off seeding and allow more mismatches and two gap open events, rather than just a single gap (see Note 11).

Use the following two commands to generate index structures for BWA and samtools:

```
quality$ bwa index strepsirrhini.fasta
[bwa_index] Pack FASTA... 0.00 sec
[bwa_index] Reverse the packed sequence... 0.00 sec
[bwa_index] Construct BWT for the packed sequence...
[bwa_index] 0.03 seconds elapse.
[bwa_index]  Construct  BWT  for  the  reverse  packed
sequence...
[bwa_index] 0.03 seconds elapse.
[bwa_index] Update BWT... 0.00 sec
[bwa_index] Update reverse BWT... 0.00 sec
[bwa_index] Construct SA from BWT and Occ... 0.01 sec
[bwa_index] Construct SA from reverse BWT and Occ... 0.01
sec
quality$ samtools faidx strepsirrhini.fasta
```

Next, we will separate single read and paired-end data (while discarding incomplete paired-end read pairs), perform alignment with BWA and convert the output to BAM:

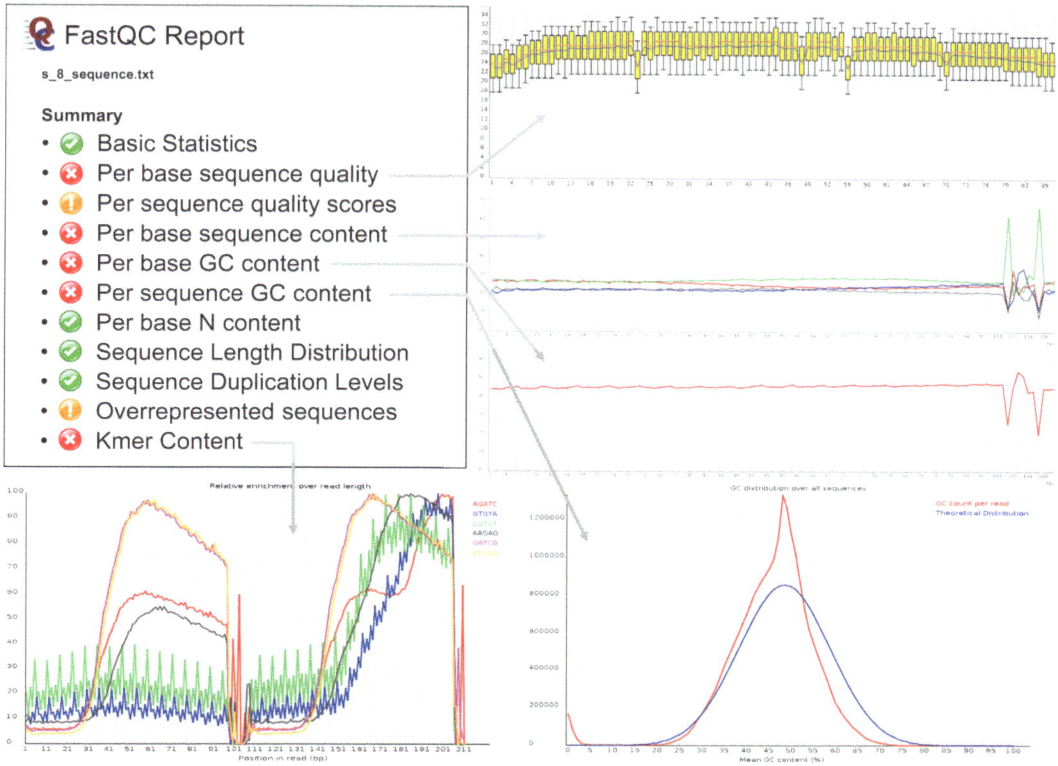

Fig. 6. Pieces of the HTML report for which FastQC reports an unexpected result. The first three plots have been abbreviated to show only the first half of the sequencing run (up to the end of the first index). In the per base quality score plot, a few cycles show reduced average quality scores, but deviations are not extreme, and this is a reasonable result of quality score calibration. Per base sequence and GC content plots show spikes around the index reads, indicating an unbalanced base composition of the indexes selected. Further, per read GC content differs from the theoretical distribution. This may be caused by larger pieces of the adapters in the sequence reads as well as a GC bias from the enrichment process. The last figure shows overrepresentation of 5-mers over the read length. The high amplitude lines result from simple repeat sequences, which were also observed by TagDust and are an enrichment artifact. The other 5-mers are motifs from the adapter sequences and indicate early set-in of the library adapters, as is expected for short insert-size DNA.

First, create a new subfolder for the output files:

```
quality$ mkdir alignments
```

Use a for-loop to iterate over the quality-filtered files, calling the SplitMerged2Bwa script for each file. Exclude incomplete paired-end pairs generated by the quality filter, and write to the output folder created above:

```
quality$ for i in *_merged_quality.txt; do \
  SplitMerged2Bwa.py --skipIncompletePE -o alignments/ $i;
done
```

Change to the alignments folder:

```
quality$ cd alignments
```

Run BWA for the paired-end data and single read data separately, and then convert the paired-end alignments with BWA sampe to SAM and the single read data (the merged reads) with BWA samse to SAM. For the combined SAM output, use samtools to generate BAM, sorted by alignment coordinates. For further details, consult the samtools and BWA manuals.

```
alignments$ for i in *_merged_quality_r1.txt; do echo ${i/_r1.txt/}; \
  bwa aln -n 0.01 -o 2 -l 65536 strepsirrhini $i > ${i/.txt/.sai}; \
  bwa aln -n 0.01 -o 2 -l 65536 strepsirrhini ${i/_r1/_r2} > \
    ${i/_r1.txt/_r2.sai}; \
  (bwa aln -n 0.01 -o 2 -l 65536 strepsirrhini ${i/_r1/_SR} | \
     bwa samse strepsirrhini - ${i/_r1/_SR}; \
   bwa sampe strepsirrhini ${i/.txt/.sai} ${i/_r1.txt/_r2.sai} \
     $i ${i/_r1/_r2} | grep -v "^@" ) | \
  samtools view -uS - | samtools sort - ${i/_r1.txt/_sshMTs}; \
  rm -f ${i/.txt/.sai}; \
  rm -f ${i/_r1.txt/_r2.sai}; \
done
```

If desired, use the samtools flagstat command to obtain simple alignment statistics for each BAM file. Use the GNU command line tool, grep, to limit the output to report the percentage of mapped reads:

```
alignments$ for i in *_sshMTs.bam; do samtools flagstat $i | \
  grep "mapped ("; done
27 mapped (0.01%)
32 mapped (0.01%)
12331 mapped (1.34%)
[...]
```

Based on these results, the samples have on average 4.03% endogenous DNA. However, some samples have as little as 0.01% and others have up to 44.09%. Here, we chose to count the number of aligned reads including all sequences regardless of their length.

Short, merged sequences may, however, result in erroneous alignments. Consequently, we should apply a lower length cutoff for analysis.

Using the samtools *view* command and the GNU tool awk, filter for aligned sequences of at least 30 nt and recalculate the percentage aligned:

```
alignments$ for i in *_sshMTs.bam; do samtools view $i | \
  awk '{ if (length($10) >= 30) { total++; if ($3 != "*")
aln ++; } } \
       END          {           printf("%d            %d
%.2f%%\n",total,aln,aln/total*100)}'; done 314879 36 0.01%
238458 34 0.01%
920089 12978 1.41%
[...]
```

The results with the lower length cutoff are very similar: 4.25% endogenous DNA (minimum 0.01%, maximum 44.09%). Here, 50 out of 96 samples show at least 0.5% endogenous DNA.

3.6. Quantifying Contamination

Contaminant sequences are those that are identified as endogenous but do not originate from the sample itself. Contaminants may be introduced from various sources, including the burial or preservation environment, people handling the sample, and laboratory chemicals. Ideally, contaminated samples should not be used for analysis. When this is not possible, however, additional steps are required. A type of negative filtration, in which sequence reads that more closely match the putative contaminant than the target sequence are excluded, has been suggested (e.g., see ref. (62)). This approach is likely to be problematic when sequences are short and/or the evolutionary distance between the putative contaminant and the sample is small.

Contamination is particularly problematic in aDNA studies of early modern humans, Neandertals, and closely related primates. Here, the proportion of endogenous reads that can be attributed to contaminants is usually estimated by examining informative sites, i.e. sites where sequence differences are known to be fixed and distinct between the contaminant and the sample (8, 13, 28). Once the amount of contamination in a sample is known, this information can be incorporated within statistical models during data analysis. If no informative sites are known, estimates of contamination may be obtained by surveying the amount of "extra" polymorphism, such as biallelic sites in haploid sequences or triallelic sites in diploid sequences. Finally, any Y chromosomal fragments in a female sample are most likely due to the presence of contaminant male DNA in the sample (8, 63).

For larger evolutionary distances, an estimate of the proportion of sequence reads that are contaminants may be obtained by attempting to align each sequence to a set of closely related reference sequences and a set of homologous contaminant sequences, and by comparing their alignment scores. Here, semi-global alignments scoring higher for the contaminants are counted as contamination, while those scoring higher for the set of close reference sequences are considered to be endogenous. Alignments that score equally are considered to be uninformative, and are excluded from the estimate.

We will apply the latter approach to our mitochondrial data set. Use the revised Cambridge Reference Sequence (rCRS) of the human mitochondrial genome as a contaminant reference sequence, as the most likely source of contamination is people who may have handled the specimens. Add the rCRS to the set of strepsirrhine mitochondrial genomes, generating a new FASTA file containing all 15 genomes. Then generate the BWA index as described above for the new sequence set *con_strepsirrhini*, and repeat the alignment procedure:

```
alignments$  for  i  in  *_merged_quality_r1.txt;  do  echo
${i/_r1.txt/}; \
  bwa aln -n 0.01 -o 2 -l 65536 con_strepsirrhini $i >
${i/.txt/.sai};\
  bwa  aln  -n  0.01  -o  2  -l  65536  con_strepsirrhini
${i/_r1/_r2} > \
     ${i/_r1.txt/_r2.sai}; \
   (bwa  aln  -n  0.01  -o  2  -l  65536  con_strepsirrhini
${i/_r1/_SR} | \
     bwa samse con_strepsirrhini - ${i/_r1/_SR}; \
   bwa      sampe      con_strepsirrhini      ${i/.txt/.sai}
${i/_r1.txt/_r2.sai} \
     $i ${i/_r1/_r2} | grep -v "^@" ) | \
  samtools view -uS - | samtools sort - ${i/_r1.txt/_con};
\
  rm -f ${i/.txt/.sai}; \
  rm -f ${i/_r1.txt/_r2.sai}; \
done
```

All reads that align to extended reference set of 15 sequences but that did not align to the original reference set of 14 sequences are counted as contaminants. In addition, all reads that score better (require fewer edits to match the reference) in the alignment to the extended reference set are considered contaminants. The ContTestBWA script evaluates the number of alignments reported for each read and the number of edits reported for both reference sets, and returns the number of good, noninformative, and contaminant reads. Assuming a binomial distribution, confidence levels can be calculated in R.

Iterate over the output files for both alignments and direct them to ContTestBWA in SAM format (calling samtools view). The *o* parameter to ContTestBWA defines an output text file *contamination.tsv* to which the result is appended; the *n* parameter determines the name printed in the first column of the output file:

```
alignments$ for i in *sshMTs.bam; do \
  ContTestBWA.py -o contamination.tsv -n $(echo $i | cut -f
3 -d"_") \
  <( samtools view $i ) <( samtools view ${i/sshMTs/con} );
\
done
```

Print the first lines of *contamination.tsv* to screen. The result will look similar to this:

```
alignments$ head contamination.tsv
Sample  Okay    NonInf  Contam  Error
che:ar1 16      2       0       0
che:ar2 15      2       0       0
che:arc 11305   10      17      0
[...]
```

Call R, read in *contamination.tsv*, and name the resulting R data structure *tbl*. Add the contamination ratio as well as the confidence interval as columns *rate* and *conf* to *tbl*. Print *tbl* on screen and write to *contamination_conf.tsv*:

```
alignments$ R --vanilla --quiet
> tbl <- read.table("contamination.tsv",header=T,as.is=T)
> tbl$rate <- apply(tbl[,c(2,4)], 1, function(x) {
    sprintf( "%.2f%%", x[2]/(x[1]+x[2])*100 ) } )
> tbl$conf <- apply(tbl[,c(2,4)], 1, function(x) {
    conf <- binom.test(x[2],sum(x))$conf.int;
    sprintf("[%.2f%%,%.2f%%]",conf[1]*100,conf[2]*100); } )
> tbl
    Sample  Okay NonInf Contam Error  rate          conf
1  che:ar1    16      2      0     0 0.00% [0.00%,20.59%]
2  che:ar2    15      2      0     0 0.00% [0.00%,21.80%]
3  che:arc 11305     10     17     0 0.15%  [0.09%,0.24%]
4  che:euo  2102     26      0     0 0.00%  [0.00%,0.18%]
5  che:gal  3020     16      2     0 0.07%  [0.01%,0.24%]
[...]
>   write.table(tbl,   "contamination_conf.tsv",   sep="\t",
quote=F,
  row.names=F)
```

The contamination point estimates range from 0.00 to 62.50%. As the data come from six modern specimens, five ancient specimens, and one museum specimen, we can calculate the proportion of contaminant sequences by sample type. All modern specimens and the museum specimen show at most 0.5% contamination. Two ancient specimens have contamination levels less than 1%. Three ancient specimen, however, contain between 8 and 13% contaminant sequences.

3.7. Handling PCR Duplicates

Because most aDNA extracts contain very little endogenous DNA, PCR amplification is often unavoidable in HTS experiments. This may lead to problems in downstream applications, as nonrandom amplification from only few starting template molecules may limit the capacity to identify polymorphisms (or damage) by altering their frequency in the sequenced sample. In the worst case, this problem can lead to incorrect consensus sequence calls or false estimates of the proportion of particular sequences at polymorphic sites. The best way to deal with this problem is to identify and remove duplicate sequences that are the result of PCR amplification.

Independent molecules are frequently identified based on their outer alignment coordinates (8, 54, 61); however, sequence-based approaches such as clustering (64, 65) may also be used. When PCR duplicates of the same molecule have been identified (a cluster), either a representative sequence is chosen or a consensus is determined. If a representative sequence is selected, that sequence should have the lowest probability of sequencing error, i.e., the highest sum of quality scores. Both the samtools (61) and GATK (54) packages implement routines for this type of filtering. A consensus sequence, which may reduce sequencing error considerably (27), should only be calculated if the duplicate sequences are very unlikely to originate from different molecules (see Note 12). A script for calculating the consensus sequence from alignments in SAM format (FilterUniqueSAMCons.py) is available from http://bioinf.eva.mpg.de/fastqProcessing/.

Identifying PCR duplicates in either the whole library or in the alignable fraction of the data may also be of interest for determining library complexity, i.e., the total number of different molecules in that library. The number of unique molecules (u) and the number of sampled molecules (s) can be used to estimate the total number of different sequences (M) (assuming sampling with replacement, one can fit subsampled values to $u = 0 + M \cdot (1 - e^{-s/M})$). Such complexity estimates are of interest when determining the fraction of a library that has been sequenced.

For the example data set, we will use a clustering algorithm called cdhit to determine how much of the library has been sampled. This algorithm has been optimized for use with the type and quantity of data typical for 454/Roche (66). Cdhit clusters all sequences that are high-identity full-length prefixes of each other. This approach is computationally very expensive compared to approaches using alignment coordinates.

Cdhit only accepts FASTA files for the clustering step. In addition, this program requires paired-end reads to be in a single line, as they reflect the read out of a single molecule.

Use a for-loop to iterate over the quality filtered files, calling the SplitMerged2CDhit script for each line. Exclude incomplete paired-end pairs, and define the *alignments* subfolder as output directory:

```
quality$ for i in *_quality.txt; do \
  SplitMerged2CDhit.py  --skipIncompletePE  -o  alignments/
$i; done
```

Change to the *alignments* subfolder.

```
quality$ cd alignments
```

Loop over the generated FASTA files, providing each as input to cd-hit-454 and defining an output file with the ".clust" extension:

```
alignments$ for i in *_cdhit.fa; do /mnt/solexa/bin/cd-hit-
454/cd-hit-454 -i $i -o ${i/.fa/.clust}; done
```

Remove the FASTA files:

```
alignments$ rm -f *_cdhit.fa
```

Evaluate the cdhit cluster files for the number of clusters using GNU tools. Print out the average coverage, the number of singletons, the number of unique sequences, and the total number of sequences:

```
alignments$ for i in *.bak.clstr; do \
  echo -n $(echo $i| cut -f 3 -d'_')"  "; \
  cut -f 1 $i | sort | uniq -c | awk '{ val += $1; count
+=1; \
    if  ($1  ==  1)  sing  +=  1  }  END{  printf("cov:
%.2f\tsingletons: \
    %d\tuniq:  %d\ttotal:  %d\n",  val/count,sing,count,val)
}'; \
done
che:ar1    cov:  1.08   singletons:  258906   uniq:  280653
total: 304358
che:ar2    cov:  1.07   singletons:  199570   uniq:  214356
total: 230413
che:arc    cov:  1.15   singletons:  608074   uniq:  689614
total: 795011
che:euo    cov:  1.14   singletons:  628662   uniq:  707199
total: 804950
che:gal    cov:  1.12   singletons:  540349   uniq:  600666
total: 673216
[...]
```

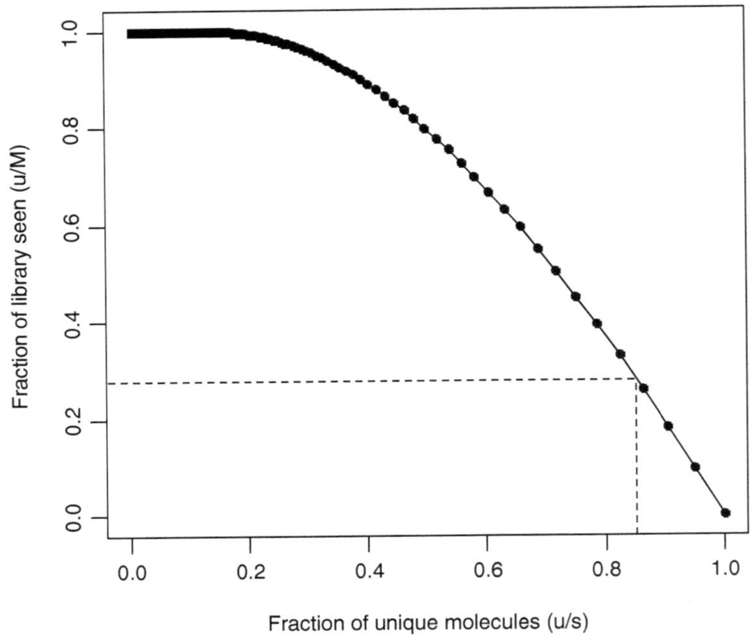

Fig. 7. Interaction between the number of unique molecules (*u*) and the fraction of the library seen (*u/M*) when assuming sampling with replacement. *Dots* show equal amounts of sampled data (*s*) being added. The number of unique molecules (*u*) relates to the total number of different sequences (*M*) and the number of sampled sequences as follows: $u = M \cdot (1 - e^{-s/M})$. For the example data set, we see on average about 85% unique molecules, indicating sampling of about 28% of the total library (*dashed lines*).

The results of this analysis indicate that the libraries have not been sequenced deeply. On average, 84.5% of the molecules are unique (minimum 29.6%, maximum 96.4%), indicating that on average 28% of the libraries were sequenced (see Fig. 7).

4. Notes

1. The FastQ sequence and quality format is not native to the 454/Roche or Ion Torrent instrument, which use a binary format called SFF (standard flowgram format). This format stores the observed flow values, and reflects that the instrument determines stretches of the same nucleotide in one go and that the number of bases in these homopolymer runs is computationally inferred. Even though SFF can be converted to FastQ (e.g., using `sff_extract` (67) or the instrument pipeline program program `sffinfo`), the conversion fixes the number of bases observed in a homopolymer run and therefore loses information. For this reason, programs working with flow values (SFF files) are preferable when using this type of data.

2. The color-space output of the Applied Biosystems/Life technologies SOLiD instrument can be converted to FastQ without performing a nucleotide sequence conversion, as the output includes four different codes for the dinucleotides. Several programs can handle color-space FastQ files (e.g., Maq/BWA (55), Bowtie (68)). Using color-space FastQ files is preferable to sequence conversion without a reference genome, as conversion may create erroneous sequence downstream of every machine error (69).
3. The Wellcome Trust Sanger Institute and Illumina Inc. use an encoding of base qualities with one character per quality scores. To achieve this, the computer representation of keyboard characters as integers is exploited. International ASCII encoding tables assign numbers 0–127 to the same characters worldwide; however, ASCII character 0–32 are nonprintable. Thus, the first character used in the Sanger standard is 33 ("!"). In its first versions, the Illumina software did not use PHRED-like probabilities (quality score = $-10 \cdot \log_{10}$(error probability)) (51). Instead, they used a quality score model with negative values, and therefore set the zero quality to 64 ("@"). In current software versions, PHRED-like probabilities are used, but depending on the exact software version still printed with the offset of 64. Other encodings include the actual quality score numbers separated by space in the quality line (e.g., AltaCyclic base call files (30)). This encoding is considered to be inefficient and is used rarely today.
4. If the sample name contains dash ("-") or underscore ("_") characters, the script will split the name by these characters and only consider the last field as the index name. The name should not contain any special characters, as it will be used as part of the name for the script's output files.
5. *K*-mers, or more specifically *n*-mers where *n* is a constant number, refers to nucleic acid or amino acid sequence "words" of a specified length. A word frequency analysis, or *k*-mer analysis, and more precisely the analysis of over- and underrepresentation of words of a specific length, can be used in different contexts. For example, *k*-mer analysis is used to identify protein-binding motifs or experimental artifacts. A set of known reference sequences can also be used to determine high-frequency *k*-mers for filtering a new data set, and the overlap of *k*-mers can be used to assemble sequences (de Bruijn graphs).
6. Sequence artifacts can be generated not only during the library preparation steps, but also during the actual sequencing process. For example, on the 454/Roche instrument, the light signal from one well position of the sequencing plate may bleed over to empty, neighboring wells and cause so-called "ghost

wells" that contain a mixture of signals. On SOLiD and the Illumina instruments, chemistry particles, dust, and lint may be identified as clusters. Frequently, these sequencing artifacts either fail the demultiplexing step because they contain an incorrect index sequence, have lower quality scores, or contain other features such as low sequence complexity that cause them to be filtered from the sequencing results.

7. The 454/Roche software automatically detects adapter sequences and marks the determined set-in point (where the non-adapter sequence ends) in the SFF data file. When exporting sequences from SFF, the trimmed sequences can be obtained. If adapter sequences are degenerated, it may be difficult to identify the set-in point, and additional measures should be taken to assure adapter sequences are recognized and trimmed. The 454/Roche software also excludes sequences for which the adapter is identified very early in the sequence read-out. The exact cutoff is version-specific, but, given the current read length from this platform, using this filter may compromise aDNA studies by causing too many short reads to be excluded from analysis.

8. The first filtration step of the raw data is generally performed by the instrument software. For example, the Illumina Genome Analyzer uses a signal purity filter called "chastity" that marks "good" sequences as PF (pass filters). This filter requires that the correct intensities for the bases called are 1.5 times higher than the next highest intensity for the first 12 cycles. In more recent versions of the analysis pipeline, the filter requires this to be true for the first 25 cycles, but allows for one outlier cycle within these 25. On the 454 platform, sequences are first filtered for a matching key sequence (the first four bases of the read) and then further filtered based on the number of active flows, i.e., how frequently a nucleotide provided for synthesis results in a base read out. Since bases are determined in a sequential order (TACG), about two active flows per cycle are expected. Reads with an average number of active flows falling at the outer edges of the distribution are frequently due to ghost wells or mixed beads.

9. The Illumina base call qualities are, like Sanger quality scores, inferred for a specific base. However, each software update provides new quality score tables and updated algorithms. On 454/Roche and Ion Torrent, the quality score concept fails with homopolymers, which are a single machine signal that is repeated over several bases when converted into sequence space. On SOLiD, quality scores incorporate a signal deduced from dinucleotides (i.e., two read outs with the chance of one of them being wrong).

10. Sequence alignments vary according to how the ends of the alignment are scored (global alignments, semi-global, and local

alignments). Widely used programs like blast (70) and blat (71) report local alignments, which find the highest scoring substring of the query sequence to a target sequence (the ends of the query sequence do not need to match the target sequence). ClustalW (72), T-Coffee (73), MUSCLE (74), and similar alignment programs typically report global alignments, where the complete query sequence is assumed to cover the whole target sequence (thus scoring incomplete coverage negatively). Semi-global alignments require the complete query to be aligned, but do not assume complete coverage of the target. It has been shown that local alignments introduce biases in the alignment of short reads to divergent reference sequences. Such biases are not observed with semi-global alignments (24).

11. Over the last several years, a number of aligners/mappers have been published with the aim to rapidly and accurately align millions of short reads (75). Some of these programs require long, perfect matches to the reference sequence (often called "seeds," see refs. (76, 77)) while others use compressed suffix arrays for efficient matching (55, 68, 78). Most programs limit the number of substitutions and insertion/deletion events to speed up the alignment. Users of the different programs should be aware of the specific limitations of each program in terms of alignment accuracy and sensitivity. Most programs are not able to consider misincorporation patterns of aDNA, which may cause them to miss alignments when the number of allowed edits is set too low. Green et al. (8) presented an aligner with sensitivity comparable to MegaBLAST (79, 80) that incorporates base misincorporation patterns typical of aDNA extracts.
12. If nonidentical sequences originating from different DNA molecules are clustered together, a consensus approach will average these. This may result in incorrect haplotype calls and low quality scores for sites where variation is present. Thus, a consensus approach should only be applied if it is very unlikely that two different template molecules may be clustered. For aDNA samples with a few million endogenous molecules, large megabase-sized genomes, and random fragment ends, the assumption of PCR duplicates as the only source is probably valid. Large amounts of endogenous DNA, small genomes, or protocols that generate nonrandom fragment ends (such as the use of restriction enzymes or multiplex PCR) may, however, conflict with this assumption.

Acknowledgments

I thank all current and previous members of the Department of Evolutionary Genetics at the Max Planck Institute for Evolutionary Anthropology, and particularly members of the aDNA and sequencing group, for interesting discussions and useful insights as well as for providing their sequencing data for analysis (especially Knut Finstermeier for providing the example data set). I also thank Knut Finstermeier and Beth Shapiro for critical reading and revisions. This work was supported by a grant from the Max Planck Society.

References

1. Margulies M et al (2005) Genome sequencing in microfabricated high-density picolitre reactors. Nature 437(7057):376–380
2. Bentley DR et al (2008) Accurate whole human genome sequencing using reversible terminator chemistry. Nature 456(7218):53–59
3. Shendure J et al (2005) Accurate multiplex polony sequencing of an evolved bacterial genome. Science 309(5741):1728–1732
4. Harris TD et al (2008) Single-molecule DNA sequencing of a viral genome. Science 320(5872):106–109
5. Drmanac R et al (2010) Human genome sequencing using unchained base reads on self-assembling DNA nanoarrays. Science 327(5961):78–81
6. Korlach J et al (2008) Selective aluminum passivation for targeted immobilization of single DNA polymerase molecules in zero-mode waveguide nanostructures. Proc Natl Acad Sci U S A 105(4):1176–1181
7. Miller W et al (2008) Sequencing the nuclear genome of the extinct woolly mammoth. Nature 456(7220):387–390
8. Green RE et al (2010) A draft sequence of the Neandertal genome. Science 328(5979): 710–722
9. Rasmussen M et al (2010) Ancient human genome sequence of an extinct Palaeo-Eskimo. Nature 463(7282):757–762
10. Krause J et al (2006) Multiplex amplification of the mammoth mitochondrial genome and the evolution of Elephantidae. Nature 439(7077): 724–727
11. Krause J et al (2010) The complete mitochondrial DNA genome of an unknown hominin from southern Siberia. Nature 464(7290): 894–897
12. Briggs AW et al (2009) Targeted retrieval and analysis of five Neandertal mtDNA genomes. Science 325(5938):318–321
13. Burbano HA et al (2010) Targeted investigation of the Neandertal genome by array-based sequence capture. Science 328(5979): 723–725
14. Poinar HN et al (2006) Metagenomics to paleogenomics: large-scale sequencing of mammoth DNA. Science 311(5759):392–394
15. Green RE et al (2008) A complete Neandertal mitochondrial genome sequence determined by high-throughput sequencing. Cell 134(3): 416–426
16. Gilbert MT et al (2008) Intraspecific phylogenetic analysis of Siberian woolly mammoths using complete mitochondrial genomes. Proc Natl Acad Sci U S A 105(24):8327–8332
17. Briggs AW et al (2007) Patterns of damage in genomic DNA sequences from a Neandertal. Proc Natl Acad Sci USA 104(37): 14616–14621
18. Heyn P et al (2010) Road blocks on paleogenomes—polymerase extension profiling reveals the frequency of blocking lesions in ancient DNA. Nucleic Acids Res 38(16):e161
19. Hofreiter M et al (2001) DNA sequences from multiple amplifications reveal artifacts induced by cytosine deamination in ancient DNA. Nucleic Acids Res 29(23):4793–4799
20. Kircher M, Kelso J (2010) High-throughput DNA sequencing—concepts and limitations. Bioessays 32(6):524–536
21. Shendure J, Ji H (2008) Next-generation DNA sequencing. Nat Biotechnol 26(10): 1135–1145
22. Reich D et al (2010) Genetic history of an archaic hominin group from Denisova Cave in Siberia. Nature 468(7327):1053–1060
23. Prüfer K et al (2010) Computational challenges in the analysis of ancient DNA. Genome Biol 11(5):R47
24. Dohm JC et al (2008) Substantial biases in ultra-short read data sets from high-throughput DNA sequencing. Nucleic Acids Res 36(16):e105
25. Lassmann T, Hayashizaki Y, Daub CO (2009) TagDust—a program to eliminate artifacts from next generation sequencing data. Bioinformatics 25(21):2839–2840
26. Briggs AW, Stenzel U, Meyer M, Krause J, Kircher M, Paabo S (2009) Removal of

deaminated cytosines and detection of in vivo methylation in ancient DNA. Nucleic Acids Res 38(6):e87
27. Krause J et al (2010) A complete mtDNA genome of an early modern human from Kostenki, Russia. Curr Biol 20(3):231–236
28. Quinlan AR et al (2008) Pyrobayes: an improved base caller for SNP discovery in pyrosequences. Nat Methods 5(2):179–181
29. Erlich Y et al (2008) Alta-Cyclic: a self-optimizing base caller for next-generation sequencing. Nat Methods 5(8):679–682
30. Kao WC, Stevens K, Song YS (2009) BayesCall: a model-based base-calling algorithm for high-throughput short-read sequencing. Genome Res 19(10):1884–1895
31. Kircher M, Stenzel U, Kelso J (2009) Improved base calling for the Illumina Genome Analyzer using machine learning strategies. Genome Biol 10(8):R83
32. Whiteford N et al (2009) Swift: primary data analysis for the Illumina Solexa sequencing platform. Bioinformatics 25(17):2194–2199
33. Noer GJ (1998) Cygwin: A free win32 porting layer for UNIX Applications. In: 2nd USENIX NT Symposium, Seattle, WA
34. Stajich JE et al (2002) The Bioperl toolkit: Perl modules for the life sciences. Genome Res 12(10):1611
35. Cock PJA et al (2009) Biopython: freely available Python tools for computational molecular biology and bioinformatics. Bioinformatics 25(11):1422
36. Mason CE et al (2010) Standardizing the next generation of bioinformatics software development with BioHDF (HDF5). Adv Exp Med Biol 680:693–700
37. Chang F et al (2008) Bigtable: a distributed storage system for structured data. ACM Trans Comput Syst (TOCS) 26(2):1–26
38. Venner J (2009) Pro Hadoop. In: Moodie M (ed) Apress. Springer, New York
39. Meyer M, Kircher M (2010) Illumina sequencing library preparation for highly multiplexed target capture and sequencing. Cold Spring Harb Protoc 2010(6):pdb.prot5448. doi:10.1101/pdb.prot5448
40. Meyer M, Stenzel U, Hofreiter M (2008) Parallel tagged sequencing on the 454 platform. Nat Protoc 3(2):267–278
41. Illumina Inc. (2008) Multiplexed sequencing with the Illumina Genome Analyzer System [PDF] [cited; 770-2008-011]. Available from: http://www.illumina.com/Documents/products/datasheets/datasheet_sequencing_multiplex.pdf
42. Stiller M et al (2009) Direct multiplex sequencing (DMPS)—a novel method for targeted high-throughput sequencing of ancient and highly degraded DNA. Genome Res 19(10):1843–1848
43. Paabo S, Irwin DM, Wilson AC (1990) DNA damage promotes jumping between templates during enzymatic amplification. J Biol Chem 265(8):4718–4721
44. Lahr DJ, Katz LA (2009) Reducing the impact of PCR-mediated recombination in molecular evolution and environmental studies using a new-generation high-fidelity DNA polymerase. Biotechniques 47(4):857–866
45. Meyerhans A, Vartanian JP, Wain-Hobson S (1990) DNA recombination during PCR. Nucleic Acids Res 18(7):1687–1691
46. Odelberg SJ et al (1995) Template-switching during DNA synthesis by *Thermus aquaticus* DNA polymerase I. Nucleic Acids Res 23(11):2049–2057
47. Mamanova L et al (2010) Target-enrichment strategies for next-generation sequencing. Nat Methods 7(2):111–118
48. R Development Core Team (2010) R: a language and environment for statistical computing. R Foundation for Statistical Computing, Vienna, Austria
49. Ewing B, Green P (1998) Base-calling of automated sequencer traces using phred. II. Error probabilities. Genome Res 8(3):186–194
50. Dolan PC, Denver DR (2008) TileQC: a system for tile-based quality control of Solexa data. BMC Bioinformatics 9:250
51. Andrews S (2010) FastQC: a quality control tool for high throughput sequence data
52. McKenna A et al (2010) The Genome Analysis Toolkit: a MapReduce framework for analyzing next-generation DNA sequencing data. Genome Res 20(9):1297–1303
53. Li H, Durbin R (2009) Fast and accurate short read alignment with Burrows-Wheeler transform. Bioinformatics 25(14):1754–1760
54. Palmer LE et al (2010) Improving de novo sequence assembly using machine learning and comparative genomics for overlap correction. BMC Bioinformatics 11:33
55. Zerbino DR, Birney E (2008) Velvet: algorithms for de novo short read assembly using de Bruijn graphs. Genome Res 18(5):821–829
56. Birol I et al (2009) De novo transcriptome assembly with ABySS. Bioinformatics 25(21):2872–2877
57. Chaisson MJ, Brinza D, Pevzner PA (2009) De novo fragment assembly with short mate-paired

reads: does the read length matter? Genome Res 19(2):336–346

58. Jeck WR et al (2007) Extending assembly of short DNA sequences to handle error. Bioinformatics 23(21):2942–2944
59. Li H et al (2009) The Sequence Alignment/Map format and SAMtools. Bioinformatics 25(16):2078–2079
60. Creighton CJ, Reid JG, Gunaratne PH (2009) Expression profiling of microRNAs by deep sequencing. Brief Bioinform 10(5):490–497
61. Green RE et al (2009) The Neandertal genome and ancient DNA authenticity. EMBO J 28(17):2494–2502
62. Edgar RC (2010) Search and clustering orders of magnitude faster than BLAST. Bioinformatics 26(19):2460–2461
63. Li W, Godzik A (2006) Cd-hit: a fast program for clustering and comparing large sets of protein or nucleotide sequences. Bioinformatics 22(13):1658–1659
64. Niu B et al (2010) Artificial and natural duplicates in pyrosequencing reads of metagenomic data. BMC Bioinformatics 11:187
65. Blanca J, Chevreux B (2010) sff_extract. http://bioinf.comav.upv.es/sff_extract/index
66. Langmead B et al (2009) Ultrafast and memory-efficient alignment of short DNA sequences to the human genome. Genome Biol 10(3):R25
67. Applied Biosystems (2008) A theoretical understanding of 2 base color codes and its application to annotation, error detection, and error correction. In: White Paper SOLiD™ System Volume. Life Technologies, Carlsbad
68. Altschul SF et al (1990) Basic local alignment search tool. J Mol Biol 215(3):403–410
69. Kent WJ (2002) BLAT—the BLAST-like alignment tool. Genome Res 12(4):656–664
70. Thompson JD, Higgins DG, Gibson TJ (1994) CLUSTAL W: improving the sensitivity of progressive multiple sequence alignment through sequence weighting, position-specific gap penalties and weight matrix choice. Nucleic Acids Res 22(22):4673–4680
71. Notredame C, Higgins DG, Heringa J (2000) T-Coffee: a novel method for fast and accurate multiple sequence alignment. J Mol Biol 302(1):205–217
72. Edgar RC (2004) MUSCLE: multiple sequence alignment with high accuracy and high throughput. Nucleic Acids Res 32(5):1792–1797
73. Trapnell C, Salzberg SL (2009) How to map billions of short reads onto genomes. Nat Biotechnol 27(5):455–457
74. Li R et al (2008) SOAP: short oligonucleotide alignment program. Bioinformatics 24(5):713–714
75. Smith AD, Xuan Z, Zhang MQ (2008) Using quality scores and longer reads improves accuracy of Solexa read mapping. BMC Bioinformatics 9:128
76. Li R et al (2009) SOAP2: an improved ultrafast tool for short read alignment. Bioinformatics 25(15):1966–1967
77. Zhang Z et al (2000) A greedy algorithm for aligning DNA sequences. J Comput Biol 7(1–2):203–214
78. Morgulis A et al (2008) Database indexing for production MegaBLAST searches. Bioinformatics 24(16):1757–1764

Chapter 24

Phylogenetic Analysis of Ancient DNA using *BEAST*

Simon Y.W. Ho

Abstract

Under exceptional circumstances, it is possible to obtain DNA sequences from samples that are up to hundreds of thousands of years old. These data provide an opportunity to look directly at past genetic diversity, to trace the evolutionary process through time, and to infer demographic and phylogeographic trends. Ancient DNA (aDNA) data sets have some degree of intrinsic temporal structure because the sequences have been obtained from samples of different ages. When analyzing these data sets, it is usually necessary to take the sampling times into account. A number of phylogenetic methods have been designed with this purpose in mind. Here I describe the steps involved in Bayesian phylogenetic analysis of aDNA data. I outline a procedure that can be used to co-estimate the genealogical relationships, mutation rate, evolutionary timescale, and demographic history of the study species in a single analytical framework. A number of modifications to the methodology can be made in order to deal with complicating factors such as postmortem damage, sequences from undated samples, and data sets with low information content.

Key words: Heterochronous sequences, Post-mortem damage, Mutation rate, Bayesian analysis, Coalescent, Demographic reconstruction, Skyline plot

1. Introduction

Nucleic acids are able to survive for hundreds of thousands of years if preservational conditions are highly favorable. Sequences of these ancient DNA (aDNA) molecules can provide a useful source of data for a variety of studies, ranging from evolutionary biology to forensic archaeology (1, 2). By applying phylogenetic methods to aDNA sequence data, it is possible to estimate the evolutionary relationships of extinct species and identify samples of uncertain taxonomic affinity. Within species, aDNA analysis can improve our capacity to estimate demographic history, past phylogeographic patterns, and evolutionary timescales (3).

Beth Shapiro and Michael Hofreiter (eds.), *Ancient DNA: Methods and Protocols*, Methods in Molecular Biology, vol. 840, DOI 10.1007/978-1-61779-516-9_24, © Springer Science+Business Media, LLC 2012

In studies conducted at higher taxonomic levels, such as comparisons among species, differences in sampling times are usually trivial in relation to the overall depth of the phylogeny. For example, the sampling times of most woolly mammoth sequences are less than 1% of the age of the family Elephantidae (4). In these instances, ancient and modern sequences can be treated as practically coetaneous and standard phylogenetic methods can be employed, including those based on the molecular clock (5, 6). At the intraspecific level, however, the sampling times of the data often span a consequential proportion of the overall history of the population—that is, the population is considered to be "measurably evolving" (7). If this is the case, failure to take into account the ages of the sequences in the data set can lead to estimation biases.

This article will focus on the analysis of samples drawn from measurably evolving populations, which requires phylogenetic methods that have been explicitly designed to accommodate heterochronous sequences. Such methods are available in a number of computer programs, including *Serial SimCoal* (8), *PAML* (9), *TREBLE* (10), and *BEAST* (11). Some of these employ a Bayesian statistical approach in which all parameters (including the tree) have a prior distribution that is altered by the observed data to produce a posterior distribution (12). The procedure described below uses the Bayesian phylogenetic software *BEAST* (*Bayesian Evolutionary Analysis by Sampling Trees*), which is able to implement a wide range of evolutionary models. Most of these models have a basis in coalescent theory (13), a statistical framework that describes the relationship between the genealogy and the demographic history of the sampled individuals.

Phylogenetic analysis using *BEAST* involves a number of discrete steps. Beginning with an alignment of the DNA sequences, an input file for *BEAST* is created using the software *BEAUti* (*Bayesian Evolutionary Analysis Utility*), available as part of the *BEAST* package. The user needs to select appropriate evolutionary models for the analysis. *BEAST* then analyzes the data set using an approach based on Markov chain Monte Carlo simulation (14). After the *BEAST* analysis is complete, the results are processed using associated software, including *Tracer* (15) and *TreeAnnotator*.

2. Materials

2.1. Data Set

DNA sequence data can be obtained using a range of methods, including those based on Sanger sequencing and pyrosequencing (*see other chapters, ibid.*). There are several important factors to consider when selecting markers for a phylogenetic analysis. Above all, the sequences need to be sufficiently variable for analysis using a Bayesian phylogenetic approach. Accordingly, a guiding principle of sampling design is to identify markers that maximize information content relative to sequencing effort.

It is important to use markers that meet the assumptions of the available phylogenetic methods. The intraspecific evolutionary models in *BEAST* are based on a relatively simple form of the coalescent, which involves the assumption that there is random mating among individuals in the study population. It is also assumed that sequences are evolving neutrally, without recombination, and without lateral transfer.

It is usually of interest to attach a real timescale to the phylogenetic estimate, which can be done by including independent calibrating information. In aDNA data sets, the known ages of the sequences can be used for calibration (16, 17). If the sampling times are unknown, they can be estimated radiometrically, stratigraphically, or phylogenetically (18, 19) (see Note 1). If the age range of the sequences spans a large proportion of the total evolutionary history of the study population, the sampling times can provide sufficient calibrating information for the analysis (7, 17).

Once the sequences are obtained, they need to be assembled and aligned. Automated sequence alignment can be performed using a number of computer programs. In some instances, there are few or no indels and alignment is trivial. Generally, however, alignment needs to be performed carefully because it can have a considerable impact on subsequent analyses.

Sequence alignments should be given in Nexus format. If the user is planning to partition the data set to allow different evolutionary models to be applied to different regions (e.g., different genes), a separate Nexus file should be created for each partition. Optional metadata, such as sampling time, can be included in the sequence name (e.g., as a suffix). If there are multiple alignment files, care should be taken to ensure consistency in sequence names across the data sets. A simplified example of a Nexus-formatted alignment is shown in Fig. 1.

2.2. Software

The analysis requires a number of different computer programs, all of which are available on the official *BEAST* website (http://beast.bio.ed.ac.uk/). The first four programs below are included in the *BEAST* package.

1. *BEAUti*. This program is used to create XML-formatted input files for *BEAST*.
2. *BEAST*. This program performs Bayesian phylogenetic analysis.
3. *LogCombiner*. This program is used to process some of the output from *BEAST*.
4. *TreeAnnotator*. This program is used to process some of the output from *BEAST*.

```
#NEXUS

Begin DATA;
    Dimensions ntax=11 nchar=1060;
    Format datatype=NUCLEOTIDE gap=- missing=?;
    Matrix

    DQ316067_32850 CAAGCAAATCAACCCGCTATGTACATCGTGCATTAAATGCTCGT ...
    EU153445_35800 CAAACAAATCAACCCGCTATGTACATCGTGCATTAAATGCTCGT ...
    EU153447_17125 CAAACAAATCAACCCGCTATGTACATCGTGCATTAAATGCTCGT ...
    EU153449_20380 CAAACAAATCAACCCGCTATGTACATCGTGCATTAAATGCTCGT ...
    EU153452_50200 CAAGCAAATCAACCCGCTATGTACATCGTGCATTAAATGCTCGT ...
    EU153454_24740 CAAGCAAATCAACCCGCTATGTACATCGTGCATTAAATGCTCGT ...
    EU153455_20620 CAAGCAAATCAACCCGCTATGTACATCGTGCATTAAATGCTCGT ...
    EU153456_18545 CAAGCAAATCAACCCGCTATGTACATCGTGCATTAAATGCTCGT ...
    EU153458_46900 CAAGCAAATCAACCCGCTATGTACATCGTGCATTAAATGCTCGT ...
    EU155210_27740 CAAACAAATCAACCCGCTATGTACATCGTGCATTAAATGCTCGT ...
    NC007596_12170 CAAACAAATCAACCCGCTATGTATATCGTGCATTAAATGCTCGT ...

;
End;
```

Fig. 1. A simple example of an alignment in Nexus format. The alignment comprises sequences from the mitochondrial control region of 11 woolly mammoths (*Mammuthus primigenius*). Each sequence name contains the GenBank accession number followed by the age of the sequence (in years), with the two fields separated by an underscore.

5. *Tracer*. This is a diagnostic program that is used to examine the output from *BEAST*.
6. *FigTree*. This is a tree-viewing program that can be used to display the phylogenetic estimates produced in a *BEAST* analysis.

3. Methods

Performing a Bayesian phylogenetic analysis can be a complicated procedure. The software *BEAST* provides a very flexible framework for implementing a variety of models, but it is for this reason that the program requires detailed input files. Fortunately, these can be readily created using the companion software *BEAUti* (Fig. 2). Once the *BEAST* analysis is complete, the output files need to be processed using further software.

3.1. Setting Up the Input File

1. Run the software *BEAUti* and import the sequence alignment(s). Details of each data set will be displayed in the window.
2. In "Data Partitions," the user can choose to link or unlink substitution models, clock models, and trees across data partitions.

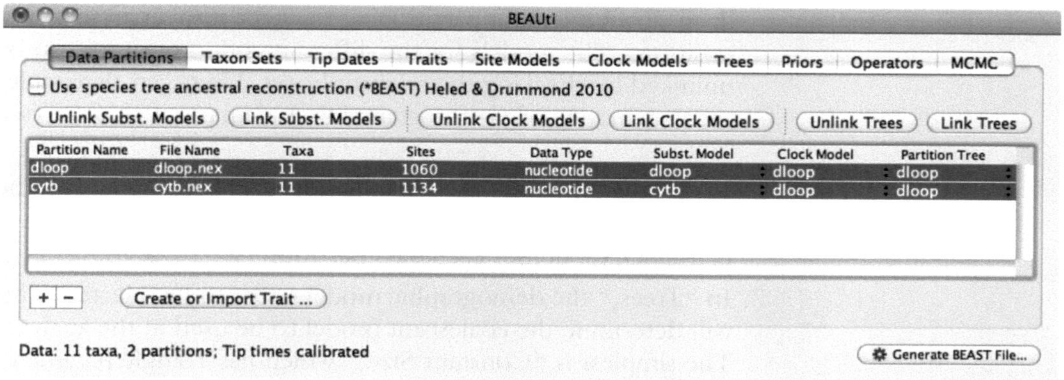

Fig. 2. Screenshot of the software *BEAUti*, which is used to create input files for the Bayesian phylogenetic software *BEAST*. The "Data Partitions" tab shows that two sequence alignments (mitochondrial D-loop and cytochrome *b*) have been loaded. The two data partitions share the same clock model and tree, but have been assigned separate models of nucleotide substitution.

If models are "unlinked," a distinct model can be chosen for each partition and parameters will be estimated separately.

3. In "Taxon Sets," define any clades that are of interest in the analysis.
   (a) Define any clades that are to be used for internal-node calibrations. If an estimate of the age of an internal node is available from an independent source, such as the fossil record or biogeography, this information can be incorporated into the analysis (see step 8 below).
   (b) Define any clades for which monophyly is to be enforced.
4. In "Tip Dates," the sampling times need to be specified for all sequences in the data set. Check the box next to "Use tip dates," select "Guess Dates," then select the option that describes the position of the sequence age in the sequence name. For example, if the age of each sequence has been appended as a suffix (as in Fig. 1), select "last" in the drop-down menu next to "Defined by its order." The dates will automatically be entered into the "Date" column. In the drop-down menu, select either "Since some time in the past" or "Before the present" as appropriate. If any sampling times are unknown, "Tip date sampling" can be activated (see Note 1). For some data sets, the sampling times might not provide sufficient calibrating information (see Note 2).
5. In "Site Models," select the optimal substitution model for each data partition (see Note 3). If a partition represents an in-frame protein-coding gene, options are available in the

drop-down menu for partitioning the gene into separate codon positions. The models for the different codon positons can be unlinked by checking the relevant boxes. Users can also implement a model of postmortem DNA damage if desired (see Note 4).

6. In "Clock Models," the default settings should generally be acceptable. Users have the option of selecting a relaxed molecular-clock model for each data partition (see Note 5).
7. In "Trees," the demographic model needs to be selected. This will determine the coalescent model to be used in the analysis. The simplest is "Constant Size," which has a single parameter describing the effective population size. In addition to simple parametric models, there are several more flexible models available (see Note 6). Users should repeat the analysis using a number of candidate models, and the results can be compared using Bayes factors (see Note 7).
8. In "Priors" the default settings can be retained unless the user wishes to include a known mutation rate or internal-node calibrations in the analysis. This can be done by choosing the form and parameters of the prior distribution for the mutation rate or for the age of a selected internal node(s) in the tree. For calibrations, a number of options are available and the choice will depend on the nature of the calibrating information (see Note 8). The internal node(s) should have been defined in the "Taxon Sets" tab (see step 2).
9. In "MCMC," the details of the Markov chain Monte Carlo (MCMC) analysis need to be specified. MCMC is a technique used to estimate the posterior distributions of parameters in the analysis, including the tree topology. The default settings (samples drawn every 1,000 steps over a total of 10,000,000 steps) might suffice for relatively simple analyses of small data sets. However, a larger number of steps will normally be required for analyses involving parameter-rich models and/or larger data sets (see Note 9).
10. Click on "Generate BEAST file" to create an XML-formatted input file for *BEAST*.

3.2. Bayesian Phylogenetic Analysis

1. Run *BEAST* using the input file produced by *BEAUti* in the procedure described above. If the data set is very large, it might be necessary to increase the memory allocation (see Note 10).
2. The output is written to two main files. Posterior samples of parameters are written to the ".log" file, while posterior samples of trees are written to the ".trees" file.

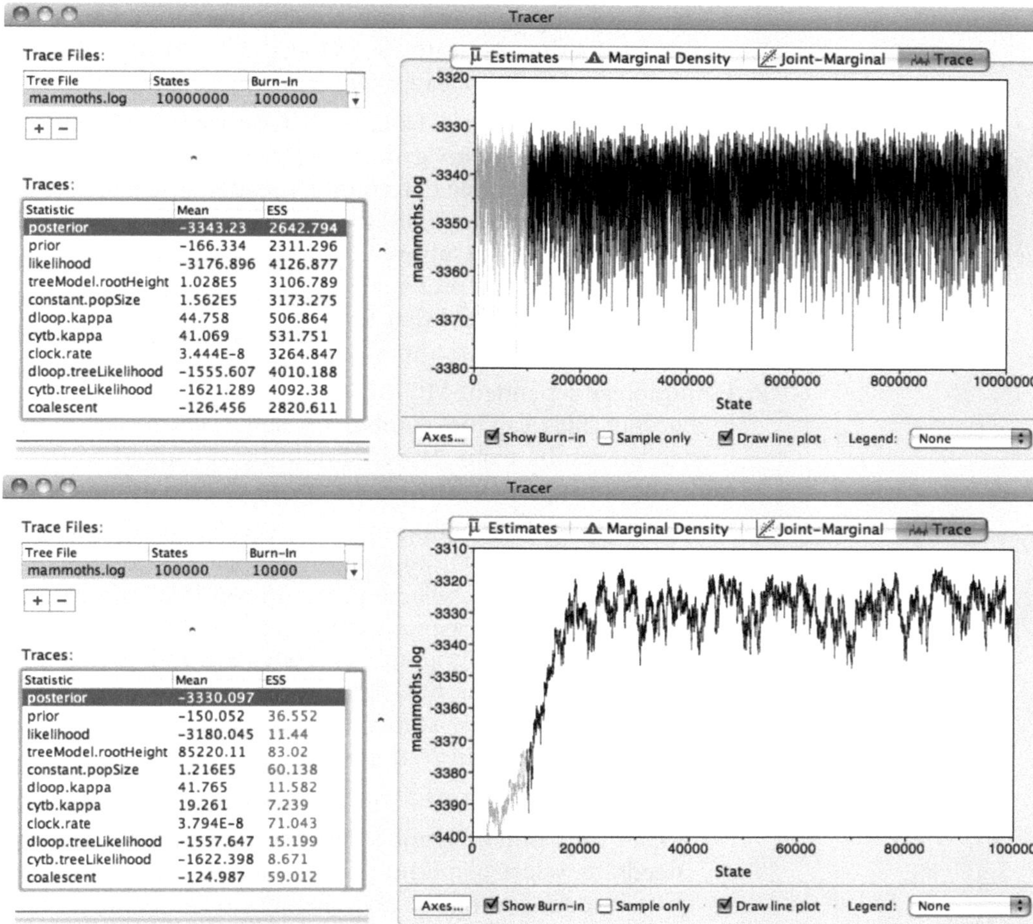

Fig. 3. Screenshot showing examples of trace analyses in the diagnostic software *Tracer*, which is used to process the output from a Bayesian phylogenetic analysis conducted in *BEAST*. In both panels, the first 10% of MCMC samples have been discarded as "burn-in" by default. The upper panel displays a trace plot from an analysis that satisfies two major diagnostic criteria. First, the plot is approximately horizontal, suggesting convergence to the stationary distribution; however, this needs to be confirmed by conducting multiple runs. Second, the effective sample size (ESS) values are all greater than 200, indicating sufficient sampling from the stationary distribution. The lower panel shows an analysis that does not satisfy the two major diagnostic criteria. First, an insufficient number of samples have been discarded as burn-in, as indicated by the upward trend in the posterior probability at the beginning of the trace plot. Second, all of the ESS values are well below 200 (shown in *red*), indicating insufficient sampling. These problems can be rectified by running the analysis for a greater number of steps and by discarding a larger proportion of the samples as burn-in.

3.3. Processing and Interpreting Output

1. Open *Tracer* and import the ".log" file.
2. Two key aspects of the MCMC analysis need to be checked (Fig. 3).
   (a) Convergence of the Markov chain to the stationary distribution is essential. This is suggested by a flattening of the

traces of all parameter estimates, but should be confirmed by conducting replicate *BEAST* analyses and checking that the traces of all replicates converge on the same values.

(b) Acceptable sampling from the stationary distribution is desirable in order to gain reliable estimates of parameter variance. This can be determined through inspection of the effective sample size (ESS) for each parameter (see Note 9). In general, values above 200 are acceptable, although higher values are preferable especially for parameters of interest. It is essential that ESS values for important parameters, such as the likelihood, are greater than 200.

3. If multiple independent MCMC analyses have been performed, the output files can be combined using *LogCombiner*. The diagnostic results from *Tracer* should be consulted to determine the appropriate number of steps to discard as "burn-in."
4. If one of the flexible demographic models has been employed (skyline or skyride plot), population history can be plotted in *Tracer*. This is done by selecting the appropriate option in the "Analysis" menu and opening the corresponding ".trees" file. *Tracer* displays the demographic reconstruction in a new window. The raw data for the plot can be exported for further analysis using statistical software or to produce publication-quality figures using illustrating software.
5. The ".trees" file is processed in two steps.

(a) Open *TreeAnnotator* and import the ".trees" file. The user needs to select a method for summarizing the sampled trees and for scaling the node heights in the summary tree. The standard method is to select the tree with the maximum product of clade credibilities and to scale the node heights to their mean posterior values. The user also needs to select an appropriate number of samples to discard as burn-in. Note here that the number of *samples* needs to be specified, not the number of *MCMC steps*.

(b) View the *TreeAnnotator* output file using a tree-viewing program, such as *FigTree*. For an example, see Fig. 4.

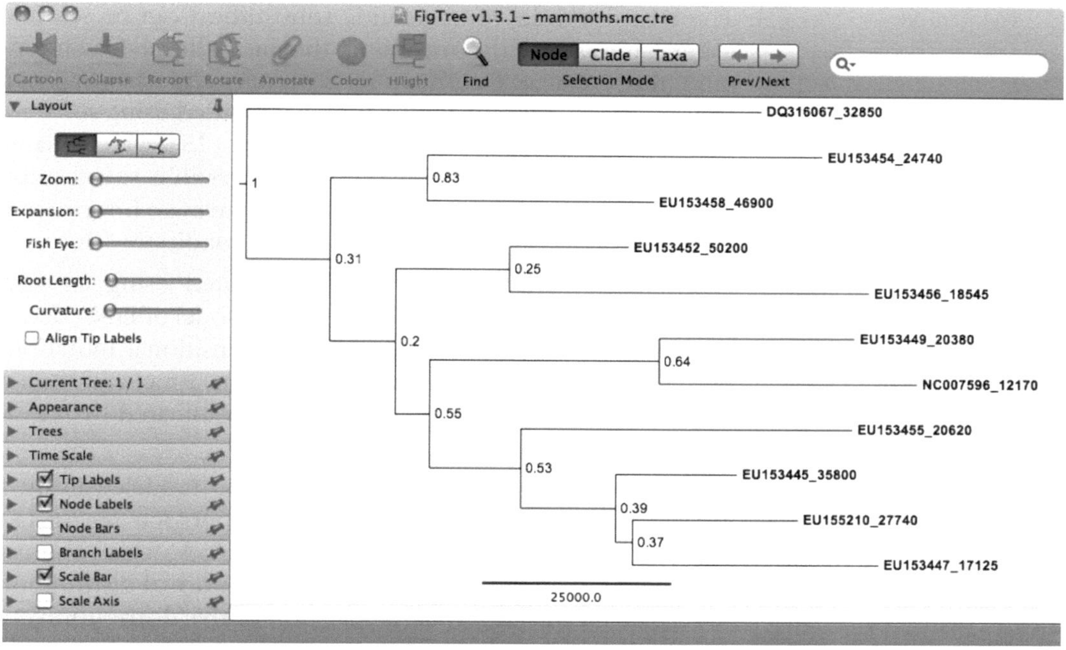

Fig. 4. Screenshot showing a maximum-clade-credibility tree displayed in the tree-viewing software *FigTree*. Internal nodes have been labeled with posterior probabilities. The tree is drawn to a timescale, with the scale bar representing 25,000 years.

4. Notes

1. Sometimes the age of a sample cannot be determined precisely by radiometric or stratigraphic means. For example, a sample might be too old to be dated reliably using radiocarbon techniques, which have an upper limit of around 50,000 years. In these cases, it is possible to treat the age of the sequence as an estimable parameter in a phylogenetic analysis (18). This approach can be taken if the sampling time is entirely unknown or if one wishes to model uncertainty using a parametric distribution (19).
2. The sampling times of the sequences are not always able to provide sufficient temporal information to calibrate the analysis. This can be investigated using a date-randomization test, which involves repeating the analysis a number of times, with the sequence ages randomly shuffled, to generate a null distribution. If the 95% HPDs of the rate estimates from the replicates all exclude the mean posterior estimate from the original data set, the sampling times can be taken as being sufficiently informative for calibration. This diagnostic approach has been used in a number of studies of aDNA (20–23).

3. The best-fit model of nucleotide substitution can be selected using various criteria, including the hierarchical likelihood-ratio test, Akaike Information Criterion, and the Bayesian Information Criterion. These can be computed using software such as *Modelgenerator* (24) and *ModelTest* (25), which enable the comparison of 56 different time-reversible substitution models. It has been demonstrated that the Bayesian Information Criterion performs well across a range of conditions (26).

4. Postmortem damage can be accommodated using an age-dependent (27) or age-independent (28) model of DNA decay. These models can be applied either to transitional mutations only, or to both transversions and transitions. Empirical studies have shown that postmortem damage is dominated by transitions (29, 30).

5. Rate heterogeneity among lineages can be modeled using a relaxed molecular clock. The standard approach in *BEAST* involves the uncorrelated lognormal relaxed clock, which assumes that the rates among branches follow a lognormal distribution (31). The mean and standard deviation of this distribution are estimated in the analysis. The model is termed "uncorrelated" because it does not involve an *a priori* assumption that rates are correlated between adjacent branches. In some respects, this model is suitable for intraspecific data because there is no expectation that rates vary among conspecific lineages in an autocorrelated manner (32). However, a more important consideration relates to whether it is appropriate to use a relaxed-clock model for population-level data at all, given that it is likely to lead to overparameterization. Drummond and Suchard (33) recently introduced random local clocks, which might be more suitable for low-information data sets that do not conform to a strict molecular clock.

6. The demographic history of a species cannot always be satisfactorily described by a simple parametric model such as exponential or logistic growth. Instead, it is possible to employ a more flexible model such as the Bayesian skyline (34) or Bayesian skyride (35). Although these methods still involve the typical coalescent assumptions of random mating and selective neutrality, they allow changes in population size to be reconstructed from the sequence data. The skyline and skyride methods can be chosen in *BEAUti* and demographic plots can be generated in *Tracer*. For a recent review of these methods, see Ho and Shapiro (36).

7. Bayesian model selection can be performed by calculating Bayes factors. The Bayes factor represents the support for one model over another, and can be calculated in *Tracer*. First, a

separate analysis needs to be performed using each of the models to be compared. The ".log" files are then opened in *Tracer*, which can be used to compute the Bayes factor for each pairwise comparison. The method used to calculate the Bayes factor, which involves computation of the harmonic mean of the log likelihood (37), has been deprecated on statistical grounds (38, 39). Better methods for computing Bayes factors have been developed, but are not yet available in *Tracer*.

8. In some cases, independent calibrating information might be available for at least one internal node in the tree. For example, the fossil record might be able to provide an estimate of the age of the root. This calibrating information can be incorporated in the form of a prior distribution for the age of the node. Several distributions are available; the choice among these should be determined by the nature of the calibration and its associated uncertainty (19, 40, 41).
9. It is difficult to determine the appropriate settings for the MCMC analysis in advance. Generally, the aim is to draw a sufficient number of samples in the smallest number of steps, but to draw them infrequently enough so that successive samples are reasonably independent of each other. The default settings in BEAUti include a total sampling period of 10,000,000 steps, with samples drawn every 1,000 steps. For analyses involving large data sets or parameter-rich models, the length of the MCMC should be increased, perhaps to 50,000,000 or more steps. The sampling frequency should be reduced correspondingly so that the sizes of the output files remain manageable. Sufficient sampling can be gauged by the ESS for each parameter, as computed in *Tracer*. ESS values can be raised by increasing the number of samples drawn from the MCMC (by increasing the total number of steps and/or decreasing the interval between successive samples), but will be reduced by autocorrelation between successive samples.
10. For very large data sets, some of the programs will encounter memory problems. This is indicated by the Java error message "OutOfMemoryError." Step-by-step instructions on how to increase memory allocation are available on the official *BEAST* site (http://beast.bio.ed.ac.uk/).

Acknowledgments

SYWH is supported by the Australian Research Council and by a start-up grant from the University of Sydney. Beth Shapiro provided helpful comments.

References

1. Pääbo S et al (2004) Genetic analyses from ancient DNA. Annu Rev Genet 38:645–679
2. Ho SYW, Gilbert MTP (2010) Ancient mitogenomics. Mitochondrion 10:1–11
3. Hofreiter M (2008) Long DNA sequences and large data sets: investigating the Quaternary via ancient DNA. Quat Sci Rev 27:2586–2592
4. Rogaev EI et al (2006) Complete mitochondrial genome and phylogeny of Pleistocene mammoth *Mammuthus primigenius*. PLoS Biol 4:e73
5. Willerslev E et al (2009) Analysis of complete mitochondrial genomes from extinct and extant rhinoceroses reveals lack of phylogenetic resolution. BMC Evol Biol 9:95
6. Rohland N et al (2007) Proboscidean mitogenomics: chronology and mode of elephant evolution using mastodon as outgroup. PLoS Biol 5:e207
7. Drummond AJ et al (2003) Measurably evolving populations. Trends Ecol Evol 18:481–488
8. Anderson CN et al (2005) Serial SimCoal: a population genetics model for data from multiple populations and points in time. Bioinformatics 21:1733–1734
9. Yang Z (2007) PAML 4: phylogenetic analysis by maximum likelihood. Mol Biol Evol 24:1586–1591
10. Yang Z et al (2007) Tree and rate estimation by local evaluation of heterochronous nucleotide data. Bioinformatics 23:169–176
11. Drummond AJ, Rambaut A (2007) BEAST: Bayesian evolutionary analysis by sampling trees. BMC Evol Biol 7:214
12. Lewis PO, Swofford DL (2001) Back to the future: Bayesian inference arrives in phylogenetics. Trends Ecol Evol 16:600–601
13. Kingman JFC (1982) The coalescent. Stoch Process Appl 13:235–248
14. Metropolis N et al (1953) Equations of state calculations for fast computing machines. J Chem Phys 21:1087–1091
15. Rambaut A, Drummond AJ (2007) Tracer, v.1.5. University of Oxford, Oxford
16. Drummond AJ et al (2002) Estimating mutation parameters, population history and genealogy simultaneously from temporally spaced sequence data. Genetics 161:1307–1320
17. Rambaut A (2000) Estimating the rate of molecular evolution: incorporating non-contemporaneous sequences into maximum likelihood phylogenies. Bioinformatics 16:395–399
18. Shapiro B et al (2010) A Bayesian method to estimate unknown sequence ages in a phylogenetic context. Mol Biol Evol 28:879–887
19. Ho SYW, Phillips MJ (2009) Accounting for calibration uncertainty in phylogenetic estimation of evolutionary divergence times. Syst Biol 58:367–380
20. Miller HC et al (2009) The evolutionary rate of tuatara revisited. Trends Genet 25:13–15
21. Ho SYW et al (2010) Bayesian estimation of substitution rates from ancient DNA sequences with low information content. Syst Biol 60:366–375
22. de Bruyn M et al (2009) Rapid response of a marine mammal species to holocene climate and habitat change. PLoS Genet 5:e1000554
23. Subramanian S et al (2009) Molecular and morphological evolution in tuatara are decoupled. Trends Genet 25:16–18
24. Keane TM et al (2006) Assessment of methods for amino acid matrix selection and their use on empirical data shows that ad hoc assumptions for choice of matrix are not justified. BMC Evol Biol 6:29
25. Posada D, Crandall KA (1998) Modeltest: testing the model of DNA substitution. Bioinformatics 14:817–818
26. Luo A et al (2010) Performance of criteria for selecting evolutionary models in phylogenetics: a comprehensive study based on simulated datasets. BMC Evol Biol 10:242
27. Rambaut A et al (2009) Accommodating the effect of ancient DNA damage on inferences of demographic histories. Mol Biol Evol 26:245–248
28. Ho SYW et al (2007) Bayesian estimation of sequence damage in ancient DNA. Mol Biol Evol 24:1416–1422
29. Binladen J et al (2006) Assessing the fidelity of ancient DNA sequences amplified from nuclear genes. Genetics 172:733–741
30. Brotherton P et al (2007) Novel high-resolution characterization of ancient DNA reveals C > U-type base modification events as the sole cause of post mortem miscoding lesions. Nucleic Acids Res 35:5717–5728
31. Drummond AJ et al (2006) Relaxed phylogenetics and dating with confidence. PLoS Biol 4:e88
32. Ho SYW (2009) An examination of phylogenetic models of substitution rate variation among lineages. Biol Lett 5:421–424
33. Drummond AJ, Suchard MA (2010) Bayesian random local clocks, or one rate to rule them all. BMC Biol 8:114
34. Drummond AJ et al (2005) Bayesian coalescent inference of past population dynamics

from molecular sequences. Mol Biol Evol 22:1185–1192
35. Minin VN, Bloomquist EW, Suchard MA (2008) Smooth skyride through a rough skyline: Bayesian coalescent-based inference of population dynamics. Mol Biol Evol 25:1459–1471
36. Ho SYW, Shapiro B (2011) Skyline-plot methods for estimating demographic history from nucleotide sequences. Mol Ecol Resour 11:423–434
37. Suchard MA, Weiss RE, Sinsheimer JS (2001) Bayesian selection of continuous-time Markov chain evolutionary models. Mol Biol Evol 18:1001–1013
38. Lartillot N, Philippe H (2006) Computing Bayes factors using thermodynamic integration. Syst Biol 55:195–207
39. Xie W et al (2011) Improving marginal likelihood estimation for Bayesian phylogenetic model selection. Syst Biol 60:150–160
40. Yang Z, Rannala B (2006) Bayesian estimation of species divergence times under a molecular clock using multiple fossil calibrations with soft bounds. Mol Biol Evol 23:212–226
41. Ho SYW (2007) Calibrating molecular estimates of substitution rates and divergence times in birds. J Avian Biol 38:409–414

Index

Beth Shapiro and Michael Hofreiter (eds.), *Ancient DNA: Methods and Protocols*, Methods in Molecular Biology, vol. 840, DOI 10.1007/978-1-61779-516-9, © Springer Science+Business Media, LLC 2012

Q

R

S

MIX
Papier aus verantwortungsvollen Quellen
Paper from responsible sources
FSC® C105338

If you have any concerns about our products,
you can contact us on
ProductSafety@springernature.com

In case Publisher is established outside the EU,
the EU authorized representative is:
Springer Nature Customer Service Center GmbH
Europaplatz 3, 69115 Heidelberg, Germany

Printed by Libri Plureos GmbH
in Hamburg, Germany